Divine Gust of Wind:

Great Trees and Their Afterlives

Other Books by Peter R. Vogt

The Monster Shark's Tooth 2009, 2010, 2020

ages 9-adult An Arizona boy visits his grandpa near the Calvert Cliffs. They canoe on the Chesapeake and through a cave into the ancient shark-infested Miocene sea.

A Most Mysterious Fossil 2011, 2020

ages 9-adult

The boy returns next year and confronts fossil thieves. He discovers a strange metal object buried in the fossil-rich Calvert Cliffs and with Grandpa's help learns the object's origin.

What Really Killed the Dinosaurs
—Evidence Found on a Montana Ranch 2022

ages 11-adult

The same boy, now a high-school grad, comes east to work as a Smithsonian museum interpreter. He meets two Montana ranch teens who invite him home to look for dinosaur fossils. The trio discover a strange object with clues (and grandpa's help) to why that asteroid killed the dinos.

Tourmaline's Quest 2022

ages 12-adult

The boy's tomboy cousin visits Grandpa to learn more about her earliest Native American ancestors. She helps him give dugout rides and falls asleep in his beached canoe. She awakes in the world of giant megafauna and sees Indian forebears. With her grandpa and cousin, she survives the comet fireballs about 13,000 years ago.

Flightship 238 2023

ages 12-adult

A now-world collector buys the diary of a Nazi-brainwashed early teenage boy. The diary details bombing and unintended involvement in a clandestine harrowing late-war escape flight on a giant flying boat. Once in Argentina, he discovers and renounces the evils of Naziism, but the diary ends. The now-world collector and his wife retrace the flight but confront dangers to learn the final destiny of teen, flying boat and cargo.

DIVINE GUST OF WIND

Great Trees and Their Afterlives

Venerating the mighty tulip poplar in life.
Reshaping it in death.

By Peter R. Vogt

Divine Gust of Wind
by Peter R. Vogt

Copyright © 2023
by Peter R. Vogt
All Rights Reserved

Editor
Sandra Olivetti Martin
New Bay Books
Fairhaven, Maryland
NewBayBooks@gmail.com

Cover by Tim Scheirer

Interior design by Suzanne Shelden
Shelden Studios
Prince Frederick, Maryland
sheldenstudios@comcast.net

A Note on type: Cover and section heads and text are set in
Garamond Premier Pro

Library of Congress
Cataloging-in-Publication Data

ISBN: 979-8-9882998-0-6

Printed in the United States of America
First Edition

DEDICATION

For my wife Randi and sons Erik and Jason—
who saw the tree fall, watched me work
on the log and rode in the dugout

Foreword

This book is about the tulip poplar—the leafy giant of eastern North America—and what I carved from one in my yard (plus two other giants) after it toppled in a storm more than forty years ago: a giant log chair; a giant round table; a giant reading nook; and a native-styled dugout canoe. Sitting in my log chair, a kubbestol, I can imagine myself as an ancient Viking. Paddling my dugout took me time traveling, and I imagined Chesapeake tidewater natives and later English colonists. Those adventures—as I canoed the hollowed-out log on the tidal Potomac, Patuxent, and Chesapeake Bay from 1983 to 2014—make up the second half of this book.

Because the tree's English names are silly and misleading, I restore the native American name, rakìock [Ra-KEE-ock], recorded by English explorer and physicist Thomas Hariot in 1585.

Table of Contents

Part I

Inspired by a Great Tree

Part II

An Ancient Vessel Returns to Chesapeake Tidewater

Part III

Splashing the Patuxent River

Introduction

My largest and favorite yard tree was now prostrate on the ground, its roots pointing at the sky, its leaves still innocently green and fluttering in the early evening breeze. This species is commonly known as yellow or tulip poplar, but as we shall see later, should have its Native American name *rakiock* restored.

Our yard's ruler had been ripped from the ground by a violent thunderstorm just before my arrival home from work. As I stared down at it from an upstairs window, I was all shock and dismay. Meanwhile my wife, Randi Vogt, was rightfully shocked and dismayed at me: Why was I not instead celebrating that she, our two young boys and our house had been spared? The tree had conveniently fallen away from the house. The next instant I came to my senses and hugged my young family, glad the tree had spared them.

Later, many thoughts flooded my mind. What to do with this fallen giant? The environmentalist in me could leave it there to return to compost, nourishing successive ecosystems of mosses, fungi, beetles and who knows what else. The tidy manicured-garden side of me could contract with a local tree service. They would butcher the arboreal corpse now lying there in state. The tree service would drown our neighborhood in chainsaw noise, irreverently hauling its remains off to some green dump, and likely making a mess of our yard in the process. Oh, and charging money I could otherwise donate to help save the Amazon rainforest. We do have a fireplace, but tulip poplar wood burns far too fast. Our region's bountiful fuelwood spoiled us, living as we do among oaks, hickories and black locusts.

For a moment, the poet in me whispered fragments from Romantic poet Percy Shelley's 1817 sonnet about the ruins of a statue to Ozymandias, a mythical ancient and long-forgotten Middle Eastern ruler. The poem describes two vast and trunkless legs of stone, with the rest of this statue a shattered, frowning visage on the desert floor. Inscribed on the statue's pedestal are the words:

> My name is Ozymandias, king of kings.
> Look at my works, ye mighty, and despair.

No matter how towering or mighty, neither kings nor trees rule forever. Our yard king had perished, as it had been born, by Mother Nature herself. Paul Bunyan never visited southern Maryland; our local good ole' boys would have turned Bunyan's Blue Ox Babe into a great bull roast. Philosophical resignation and humor made me feel better.

What about making mementos out of the wood? That is done with famous trees, such as the Wye Oak, of Talbot County, Maryland, or the Liberty Tree of Annapolis. I was, after all, an amateur wood carver. But so far only smallish things like noggins, gavels, soup ladles and butter knives. In spreadsheets with wood characteristics, no species scores a five in every column.

Tulip poplar wood is relatively soft, weak and lacking in figure. A tulip poplar gavel would break on first use by a local hanging judge.

Living for years near the Chesapeake Bay, I have picked up some of its history and prehistory. I knew that Native Americans in the Maryland Bay area hollowed out tulip poplar logs to make dugout canoes, and that English colonists learned this from the natives and expanded on the concept. In fact, a few years earlier, I had started on such a canoe but gave up. Our Ozymandias was a giant and deserved a giant project! I would carve a dugout out of the straight part of the trunk.

My thoughts about a dugout were intruded by thoughts about the tree's massive, flaring butt end. No good for the canoe. Could I leave it as a jungle gym for kids to climb up among the roots? Then I had a better idea. My wife, Randi, is Norwegian American, and we have visited Norway, where wood carving is traditional and still practiced. The Beatles' song "Norwegian Wood" rang in my head. No dugout canoes there, but giant solid wood chairs, called kubbestols. I would make one of those to celebrate my wife's ancestry.

The wind-thrown tree had started on the long adventurous path that led to this book.

PART I

Rakìock

CHAPTER I

America's Leafy Giant

The star of this book is raklock—the tallest tree in eastern North America. Of all eastern U.S. species, only white pine can compete, and only the sycamore has greater girth. Called king of the magnolia family, the raklock is the official state tree of Indiana, Kentucky and Tennessee. Fast growing and lighter (meaning less dense) than even most softwoods, this hardwood (better given its native name) remains a major lumber source, for boxes and nobler uses. raklock flowers are prized by bees, beekeepers and honey fanciers. The Native Americans of Southern Maryland and elsewhere carved dugout canoes from it, as later in this book we will see.

As a living tree, raklock abounds in virtues. No writer has lavished more praise on it than botanist-naturalist-poet Donald Culross Peattie in my favorite and most dog-eared tree book, *A Natural History of Trees of Eastern and Central North America*. This tree, he gushes, is of "stately beauty and immense practical use," glorious in spring "when its flowers, erect on every bough hold the sunshine in their cups, setting the whole giant tree alight."

Describing its form, Peattie finds "something joyous in its springing straightness, in the candle-like blaze of its sunlit flowers," its leaves on pendulous slender stalks "forever turning and rustling in the slightest breeze." Its palette of colors extends beyond the flowers "yellow or orange at the base, a light greenish shade above" to the "glossy, sunshiny pale green" leaves that "in autumn turn a rich, rejoicing gold."

In stark contrast to Peattie's poetic praise, for most of the public, rakìock is just another tree. Its tulip-like flowers are too high up to be seen. Nothing as common as this species, especially in Southern Maryland, is special. It's how we humans are—the opposite of "absence makes the heart grow fonder." Moreover, some homeowners complain about the 'litter' of leaves, sticks and seeds. And among some, the arborphobes, the rakìock is feared and hated because some have fallen on and even destroyed houses in storms.

Lacking any similar European relatives, the tree's ancient name—in the native "Virginian" language of coastal North Carolina—was recorded by English renaissance scientist Thomas Hariot in 1585 as rakìock. While that language —today termed Carolina Algonquian—is long extinct, Hariot learned it from natives so as to quiz them and serve as interpreter. He even invented an early way of writing that language phonetically. Hariot described "rakìock [as] a kind of trees so called that are sweet wood of which the inhabitans that were neere vnto vs doe commonly make their boats or Canoes in the form of trowes." Hariot placed an accent mark over the letter "i" to show it should be pronounced Ray-KEE-ock. In Hariot's time, the English language had not yet acquired a standard orthography (spelling), even in print.

The Indian name rakìock was and remains largely forgotten. Not at all a giant tulip, nor any kind of poplar, the tree was misnamed by later English colonists. These settlers were fooled by the tulip-like flowers and its low-density wood, like that of many poplars.

But rakìock acquired plenty of other English names, depending on where it grows: poplar, yellow poplar, popular, popple, tuliptree and, in pioneer days, whitewood. In the Appalachians, some called it canoe wood. This last name recalls Hariot's description of rakìock. Let this book be the first to recommend restoration of its original native name. rakìock can then join the sadly short list of tree names—notably persimmon and hickory—derived from native roots.

As to what the word rakìock originally meant, it's unlikely anyone will ever know; linguists are not even sure what Chesapeake originally meant.

Botanists avoid popular names just as the public avoids scientific ones. (T. Rex is an exception in the animal world.) There are too many popular names, mostly misleading, for the same plant. And what about languages other than English? The scientific name of rakìock is the euphonic Liriodendron tulipifera. The genus tells of lyre-shaped leaves and the species of tulip-shaped flowers. Let this scientific name roll off your tongue! rakìock had already become popular in Europe in early Colonial times. In fact, the species was described and named from a mature tree in a Leiden garden in the Netherlands in 1687.

Alas, rakìocks have acquired a bad rap because they can uproot or break in storms and crush houses. We will get to that later.

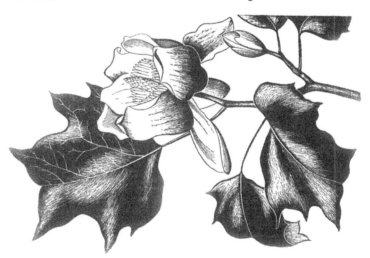

CHAPTER 2

Sacred, Historic and VERY Big Trees

Wherever we humans have lived among trees, the biggest, tallest and oldest have been revered and even worshiped. Among many others, the pagan Norse and Saxons held sacred certain great trees or groves. The Norse had their mythical Yggdrasill [EEG-drah-zil], the enormous tree sacred in their cosmology, perhaps originally a large ash or yew. An actual giant tree was venerated near the Viking temple at Uppsala, in Sweden.

In a story rich in symbolism, the English missionary St. Boniface (born Wynfrid), attempting to convert Germanic Saxon pagans, felled their sacred tree, Thonar's Oak—also known as Donar's Oak, Thor's Oak and Jupiter's Oak—about 723 CE. Allegedly, he was assisted by a "divine gust of wind." The monk and his supporters repurposed some of the wood to build a chapel. (Thor or Jupiter apparently failed to intervene.) However, even fifty years later, the Saxons had still not been Christianized and still venerated old oaks. Charlemagne sent his troops to destroy the pagan Irminsul—a large old oak adorned with icons, or a kind of totem pole column made from such a tree. In later years, entire groves of sacred oaks were felled on orders of Charlemagne. Thousands of Saxon nobles suffered the same fate as their special trees. With the establishment of Christianity, any remaining signs of pagan tree worship became criminal offenses.

The historic record of Germanic peoples goes back even farther than St. Boniface and the Irminsul. The Roman historian Tacitus (Germania, 99 CE), reported on what he learned from travelers. Northern tribes practiced their rituals not in temples but in groves of trees. When this practice first began is lost in prehistory, but perhaps it started when large forested areas were being cleared for farming, around 8,000 years ago. As remaining groves became scarce, they grew more valuable—not only for wood but as imposing symbols of longevity.

On this side of the Atlantic, there is little evidence for native peoples worshiping large trees. But who's to say some tribes didn't? Had rakìock been native to Europe, some giant ones would surely have been held sacred. In the Chesapeake region, a large rakìock was more likely valued as dugout material. In 1797 in New York state, a gigantic oak was chosen for the signing of the Treaty of Big Tree, in which The Seneca Nation ceded 3.5 million acres—most of the tribe's land—to the US government. This oak was swept away by a flood in 1857.

In 1765, a stately American elm growing near the Boston Commons became the site for protesting British taxation and a place for Colonists to rally and hang effigies of British authorities. In an echo of St. Boniface, ten years later British Loyalists cut down this prototype Liberty Tree and, to further spite the patriots, used it as firewood. However, other prominent trees were thereafter designated Liberty Trees, first in the rebelling 13 colonies and later during the French Revolution. These trees served as living symbols opposing repression.

In Maryland, one ancient rakìock became our Liberty Tree. It grew on the campus of St. John's College, in Annapolis. In another echo of the Donar's Oak of St. Boniface, the Liberty

Tree also fell victim to a gust of wind—during Hurricane Floyd in 1999. As with many hollow old trees, its 400 year age is only an estimate. The remaining sound wood of the Annapolis Liberty Tree was repurposed into various mementos, including guitars. Seeds and seedlings were distributed to high places around Maryland and in all original 13 colonies. This Liberty Tree lives on via its numerous descendants—and via the music played on its guitars.

Our present world still has monument rakìocks. The Queen's Giant, first sprouted around some 350 to 400, possibly even 450 years ago, may be the oldest rakìock living today. Queen's Giant—138 feet tall in 2005, grows in New York City in a remote corner of Alley Pond Park, just 200 feet or so from the Long Island Expressway. For some years, the tallest rakìock, 177.4 feet in 2013, grew in Great Smoky Mountains National Park. The current tallest rakìock reaches 192 feet, also in the southern Appalachians. Another giant of 168 feet growing in the Smokies contains the third highest volume of wood—243 cubic feet—of eastern US trees. With a 7-feet DBH (Diameter at Breast Height) this old-growth rakìock sports a crown 101 feet across.

What really constitutes a Big Tree? Height? Girth? The diameter of its crown? Maryland's first state forester, Fred Besley, in 1925 was the first to define Big Tree. His formula for bigness is based on a score computed by adding three measures: height in feet; circumference at breast height; and one-fourth the diameter of the crown. Although this formula is somewhat arbitrary, it has since been adopted by every state and beyond. States have Big Trees of each species within their boundaries, and many counties have their own Big Trees. As of late 2022,

the Maryland champion Big Tree of all species was a sycamore. However, at No. 3 stood a raklock. Of the top ten Maryland Big Trees, six were raklocks. The largest tree, the sycamore, totaled 499 points (145 feet high with a 326-inch girth and a 111-feet crown). The Maryland champion raklock came in with 460 points (136 feet height, 292-inch girth, and 127-foot crown. It grows in Southern Maryland, in Charles County in the unincorporated community of Bel Alton.

Ideally, tree age should be the fourth Big Tree measure. However, this requires the expertise of a forester or dendro-chronologist—a studier of tree rings—to extract a pencil thin core, then patch up the small hole to keep out pathogens. The annual rings can generally be counted in the core sample. For very large trees, the corer won't reach the oldest rings in the center, and in any case the tree may be hollow. So extrapolation is in order. The oldest living tree in Southern Maryland is probably a 5-plus feet DBH bald cypress (Taxodium distichum) safely growing in the Battle Creek Cypress Swamp Nature Sanctuary, in Calvert County. This tree sprouted well before the first English colonists arrived and is likely roughly twice as old as the Calvert Big Tree champion, a raklock. When measured in April 2014, the old bald cypress was 204 inches in circumference, 68 inches in diameter and 132 feet tall, with a 77 foot crown. That's a Big Tree score of 364 points. Whether it would beat out Calvert's Big Tree champ if age were factored in would depend on how age is weighted.

The Calvert County champ Big Tree is a raklock growing near a hiking trail in Flag Ponds Nature Park, not far from the Calvert Cliffs. In 2014 it was 138 feet tall, 20 feet, 2 inches around and 77 inches in diameter, sporting a canopy 99.5 feet across. At 408 points, that's just 44 points more than the Battle Creek bald cypress. Its hollow interior is big enough for me to stand inside and look at the sky above through a gap. Hollow and ancient, this tree has nonetheless survived any number of tropical cyclones. The age cannot be exactly determined, but surely it's 200 years or more.

Hollow trees, like those crooked and forked, had little fuelwood or lumber value, and during centuries of intensive agriculture provided shade to men and their animals. And leaning hollow trees are less likely to succumb to gravity compared to leaning solid trees. As aeronautical engineers learned by inventing tubular spars, hollow cylinders are quite strong relative to weight.

Rakìock and sycamore will likely continue competing for Maryland Big Tree champion. Not a surprise; among the many eastern hardwood species, one is the tallest and the other has the greatest girth.

We can be rather sure that today's record specimens are not all-time records. Some tree experts suggest exceptional rakìocks might have reached 200 feet in height with 8-to-10-foot diameters in old growth forests prior to European settlement. Such forest-grown trees would have had columnar trunks clear of branches for the first 80 to 100 feet. Around 1800, the surveyor general of North Carolina encountered a hollow one wherein:

> A Man had his Bed and Household Furniture
> and lived in it till his labour got him a more
> fashionable Mansion.

TREES BUTCHERED; VENERATED

Rakìock is a fast-growing species and can reach 3 to 5 feet in diameter in its first century. Fast growth often leads to overestimated age. Size is only a crude measure of age; the largest tree in our community, near the Calvert Cliffs, was also a rakìock. At 6 feet in diameter, this giant was locally believed to be at least 300 years old. After needless execution of this so-called hazard tree in 1986, its complete set of annual rings numbered 155, give or take a couple rings. Two such rakìocks could have grown in succession since the first English colonists arrived. More about that tree later.

The double line of rakìocks forming the Poplar Walk at Nomini Hall near the tidal Potomac were "only" planted in the 1750s. Being field-grown, they now sport massive lower limbs and attained 20 feet girth in just two centuries.

While there are few people today who actually worship giant trees, most of us at least venerate them as living majestic connections to times before us two-legged mayflies. People come to admire and pose for photographs either next to or for hollow ones even inside them. Laws may punish those who destroy prized, healthy trees. The most famous are fertilized, treated against pathogens, pruned and cabled. The Annapolis Liberty Tree would likely have succumbed prior to 1999 had it not been variously cabled and outfitted with steel rods, its large cavity filled with 55 tons of cement. If trees were sentient beings, this tree would have been grateful for all the efforts, albeit not for the public trampling its roots.

Centuries ago someone might be hanged for worshiping trees that would be felled by iconoclasts, as St. Boniface felled the Donar's Oak. Today, worshiping a giant rakìock or another large old tree might be considered bizarre but, hopefully, not illegal. If worshiped trees could speak, they would ask the offerings be human body waste, not gold trinkets.

If there is one other Eastern US tree species that deserves veneration, it's the beech (Fagus grandiflora). A majestic tree, so familiar to the European beech that English colonists used the same name—descended from its Anglo-Saxon name *beece*, in turn derived from boc, meaning letter or character, from which we got the word book. This connection between book and beech derives from the ancient use of the smooth-barked beech for carving inscriptions, such as names or messages, and often love poems or just 'AB + CD' inside a heart.

The linguistic roots of the word pair beech-book likely trace back to the Proto-Indo-Europeans, who spread from what is today Ukraine and southwest Russia into Western Europe some 5,000 years ago. Those people brought along their language and probably traditions of carving messages into beech trees. This explains why this word pair is similar in modern languages of northern and eastern Europe. (For example, Buche-Buch in German, bok-bok in Norwegian, and the Russian word for beech, buk.) The earliest northwest European alphabet is Runic, which by its mainly straight lines is more easily carved into tree bark or chiseled into stone, compared to rounded Latin letters.

Perhaps the most famous American message carved into a beech was that of Daniel Boone. It read 'D. Boone Cilled A Bar On Tree In Year 1760'. Boone's inscription was no longer legible when this 70-foot-tall, 28.5-foot-girth, 365-year-old beech fell down in 1916. It stood in Washington County, Tennessee. The tree outlived—by two years—the last living passenger pigeon, a species that depended on beechnuts and was once the most numerous bird in North America. Deforestation of beech trees and industrial-scale pigeon slaughter spelled doom. Not as tall or thick as rakiock, the beech deserves the title America's memorial tree—a monument to the greed and stupidity of the human species.

HISTORIC DANIEL BOON TREE, Copyrighted, 1906, and Published by
Near Jonesboro, Tenn. Reese & Read, Knoxville, Tenn.

While tree-worshiping is almost forgotten today, our modern culture has—in film and fiction—morphed them into speaking purveyors of ancient wisdom. The Ent named Treebeard in *Lord of the Rings* is an ancient shepherd of forests, a teacher of forest preservation. I imagine Treebeard as a walking, talking English oak. Another wise movie (but not moving) tree is Grandmother Willow, in *Pocahontas* (1995).

However, the real Pocahontas would never have seen this Chinese tree, except maybe in London. Moreover, such trees are short-lived. But like tree-worshipers, we movie-goers have to suspend disbelief. The real Pocahontas lived among rakìocks, rode in dugouts made from them and knew them by that or similar name. Maybe a remake will create a Grandfather Rakìock.

CHAPTER 3

Tulip Poplar Paradise: Western Shore of the Maryland Chesapeake

Lyriodendron tulipifera (aka tulip poplar or raklock) is native to most of the eastern United States, from the Mississippi east to the Atlantic. Its natural range extends southwards from southern Ontario and southern Michigan and along the East Coast as far north as southwest Massachusetts and Rhode Island. Central Florida is a southern limit, as is Louisiana. With climate change, raklock range will likely expand northwards, into southern Quebec and further into Ontario.

Nowhere is it more abundant than along the western shore of the Chesapeake, where raklock is the dominant species—unlike elsewhere in its range. This dominance was initially overlooked in the late 1970s when paleoecologist Grace Brush and her Johns Hopkins University students were mapping Maryland forests. (More about this dominance below.)

Raklock prefers rich moist soils but eschews soggy ground. With a relatively shallow root system, they do not tolerate drought. raklock tolerates soil pH from 6 to 8, so a neutral pH of 7 may well be the best. Saplings can grow upward fast, reaching 40 feet in just fifteen years. The species' shade intolerance explains why it evolved to grow fast: Its crown grabs sunlight before competing species. This also explains why young tall trees shed their lower branches, considered unsightly litter and possible widow-makers by gardeners manicuring the land below.

Compared to other species, raklock roots are fleshy, easily injured and broken.

Raklocks flower after their age passes at least fifteen years. After flowering, gravity and breezes carry the samara-like seeds as far as four to five times tree height, perhaps 500 feet. Seeds remain viable for four to seven years.

The pyramidal crowns are maintained into old age, which can reach 500 years but is commonly about 200 to 300 years in old growth forests, such as Belt Woods, a 43-acre remnant in Prince George's County near the intersection of Church Street and Central Avenue. The Belt Woods canopy species are dominated by white oak and raklock, clear evidence that the latter species can hold its own for several centuries. The common canopy trees there are typically 3 to 4 feet DBH (diameter at breast height) and 140 to 150 feet tall. Isolated raklocks can rise to at least 192 feet, presently the tallest known tree in eastern North America. That giant, dubbed the Tall One, was discovered by the Eastern Native Tree Society in 2011 in a remote corner of Smoky Mountains National Park.

The Tall One and other giants grace old-growth remnants of the Southern Appalachians. However, areas never logged are more common there; similar giants may once have grown on the Maryland Coastal Plain.

In areas where raklock dominates, it is the species that most commonly blows down in a storm. That would be expected, everything else being the same. While there may be other factors, such as their great height and relatively shallow root system, the main reason more raklocks occasionally fall or even (rarely) destroy houses is their greater abundance. In other parts of southern Maryland, where oaks dominate in developments, oaks are blown down on houses.

So the question should be, why is the western shore such a paradise for rakìock?

The soil substrate comprises shallow marine sediments, great at holding moisture and, in many places, buffered by fossil shell beds. These sediments were laid down in shallow embayments of the Atlantic during the middle parts of the Miocene Epoch. This is a gift from geology. There is also a gift from geomorphology. Atypical for the coastal plain, the terrain is incised by steep-sided ravines and stream valleys, mandating that fields for crops are small.

The principal money crop from the mid-17th to late-20th centuries was tobacco. When the tobacco fields were abandoned for suburban development, beginning with summer homes, rakìock trees growing along steep ravine margins easily spread their seeds all across small fields. Crops of even-aged rakìock emerged, their fast growth preventing other species. Some trees were removed to build houses, but the rest were left to their own devices. As true in any pioneer forest sprouting in a bare field, the forest became overcrowded (overstocked). Weaker or slower-growing trees perished. This natural thinning would continue for some decades. Root systems of close neighbors remained small, adding to risk of uprooting. Moreover, homeowners unintentionally damaged the fleshy roots via excavations of many kinds.

After fifty to seventy-five years of development, the remaining rakìocks are expanding their root systems and canopies. With proper management, they can live for another century or two.

Communities wishing to live close to nature can live close to rakìocks. To reduce sail area (wind resistance), the canopies should be pollarded at intervals. At least 25 percent and often more of a canopy can safely be removed. Dead or weak branches should be pruned off. Tree heights can be limited by selective topping. The Critical Areas prohibition of topping should be revised. Many trees survive natural topping by storms. Damage to root systems can also be avoided.

Rakìocks could even be maintained (as in bygone times) as coppices. These are woodland areas where trees are periodically cut back to stumps to promote sprouting of foliage that can be harvested to feed livestock. Stop vilifying and start managing our American rakìock, aka tulip poplar. Learn from George Washington.

Meanwhile, a basic question remains unanswered: How common were rakìocks in the natural old-growth forests first encountered by English explorers and colonists? Palynologists examining sediment cores in the Chesapeake Bay mainstream have characterized ancient forests as dominated by hickory and oak, with other species such as walnut, chestnut, pine and others. Lyriodendron tulipifera is not on these lists because it's insect pollinated; hence the fine honey. rakìock has "learned" not to waste energy on producing the massive amounts of pollen needed for wind dispersal.

I'm not the right kind of scientist, but IMHO—In My Humble Opinion—the only way to estimate ancient rakìock abundance is to count pollen from several wind-pollinated proxy species, or to comb through sediment samples for the scarce rakìock pollen.

CHAPTER 4

Divine Gust of Wind

Thunderstorms are common during the Southern Maryland warm season. Usually they pop up locally on hot, humid afternoons, with perhaps a modest gust of wind and a quarter or half- inch rain. Hail is rare. Temperatures drop slightly, but not humidity.

The first Tuesday of June, 1980, was just another workday for me in Washington DC. Our house near the Calvert Cliffs along Chesapeake Bay is about an hour's commute away. Nothing unusual on the way home. Some road areas were wet from scattered thunderstorms.

As I got closer to home, there were more and more leaves and twigs, and then small branches on a wet, hail-covered ground. There was no rain or wind. As I neared our house, I was shocked at what I saw: Our centenarian raklock—the largest tree in our yard—had been uprooted and lay pointing downslope, its massive root ball rising 10 feet in the air.

It had missed our house.

Nearby stood our great, 125-foot-high, 30-inch DBH (diameter at breast height) bitternut hickory (Carya cordiformis), half of it ripped off by the same gust. That wound was far too big to heal, and five years later, after oyster mushrooms had sprouted along the trunk, I had to have the tree removed. Another bitternut (22-inch DBH) had been uprooted and in falling clipped one corner off our roof.

We repurposed that tree for firewood. Hickories have strong deep root systems and require a truly violent gust to uproot.

Although never accurately measured for height or crown diameter, our raklock was not remotely a record for the species. Its DBH had increased from 32 inches in 1970 to 36 inches in 1980. (You have to live in the same place a long time to record the very slowly increasing girth—inches per decade—of mature canopy trees). Before death our tree's height had been about 100 to 110 feet, the crown diameter perhaps 75 feet. These measures yield a Big Tree score roughly 225 to 250, only half that of the Annapolis Liberty Tree. Our raklock did not rule state or county, but in our immediate neighborhood it was king. Moreover, due to growing on a slope too steep to till, it dated from before our community was developed from cleared tobacco farms.

This happened just moments before my arrival home. It had caught my wife and two young sons by surprise. They were standing in the living room and saw these trees go down! I was so distraught by losing these revered trees that my unhappy wife had to remind me that everyone and mostly even the house survived intact. Shame on me!

In another age, I would have wondered if this was a Great Spirit's divine punishment for my misdeeds. Or was it a gift with strings attached: Am I to make wonderful things from the wood, like the chapel St. Boniface built from Donar's Oak in CE 723? Our house was spared, so maybe the gust was just a warning to shape up.

We now live in the age of science. Meteorologists often post warnings of high winds, perhaps with gusts up to 60 miles per hour. The great majority of folks in such warning areas never experience such possible, but highly unlikely, gusts. Being a scientist, I would say our yard was in the wrong place at the wrong time. Maybe a statistical outlier.

Was there some connection with the massive eruption of Mount St. Helens just a few weeks earlier? No way to know that.

When a tree falls in areas occupied by people, what happens next depends on where the tree fell. If it or large limbs fall on houses, cars, power lines or roads, few people think about the tree, let alone pining for its loss.

If the tree damaged or even demolished a house beyond repair, or even injured or killed a person or pet, it's an emergency. The focus is on first aid or possibly calling an ambulance and then getting the roof covered with tarps before rains damage house interiors. After that, roof repair contractors would be contacted. Occupants may have to look for temporary lodging.

If the fallen tree or limbs block walkways or roads, neighbors might help cut up the tree with axes and chainsaws and pull the slash off the road. Tree companies would be called if required. Danger lurks where the fallen tree has downed a power line, requiring prompt arrival of utility companies.

Where trees fall harmlessly into yards (as in our case) there is no emergency. Where homeowners associations require downed trees or snags be removed, a tree service has to be hired. Barring such rules, environmentally conscious homeowners might leave the snag or stump and/or logs and landscape around them. I try to recommend that to neighbors. As the dead wood gradually decays, it supports successive diverse ecosystems of fungi, mosses, beetles, etc. Homeowners may even be rewarded with a crop of tasty oyster mushrooms.

Alas, the typical alternative—preferred by those many folks with a "manicured suburb" mindset—is to have the downed tree cut up and hauled to some green landfill. Tub grinders are loud, and digging a stump out of the ground and then filling the pit is messy, noisy and wasteful, requiring a higher cost in money and carbon footprint.

Downed trees can often be repurposed into firewood, with the slash (branches and twigs) turned to mulch with a power chipper. The latter machines are painfully loud and dangerous but commonly accept branches up to 4 to 6 inches in base diameter. Instead of paying to have the chips hauled away, have the chips blown into some nearby patch of woods or into a mulch pile. Upon losing a hickory to bark beetles, I had the tree removal company leave logs no longer than two feet. Then I invited neighbors to share splitter rental costs, split the logs and share the firewood. Hickory is easy to split when still green, and it makes outstanding firewood.

Rakìock has many virtues, but not as fuelwood. Barely more than half the density of hickory, it burns too fast. At least rakìock does not—as resinous conifer wood does—throw sparks or coat your chimney with flammable unburned soot.

After my sorrow had faded, I decided the shattered visage of my noble rakìock deserved a fate more memorable and permanent than humble decay or fire. Already a few years before, the ghost of Tom Sawyer had released a dugout canoe bee into my bonnet. I had suggested to our local museum that they really should have one among their many historic boats. I was on the lookout for a suitable log. There was one in our neighborhood, but after some desultory chipping it looked too small. Now, the Great Spirit had presented me with one of the right size right in my own backyard.

Before starting on the dugout, I remembered those Scandinavian log chairs called kubbestols I had seen and admired in Norway. My downed rakìock widened (flared out) a lot near its base of roots. And fluted as well. Many older trees do that. Those features are not suitable as part of a dugout, which require a straight cylinder. Then it hit me: Why not carve a giant kubbestol from this flaring base? Once that bee was buzzing in my bonnet, I decided to make the kubbestol first, then proceed with the dugout.

I had not previously fashioned any wooden objects larger than a trencher or soup ladle. Now I was embarking on liberating two enormous things from the giant log in which they were trapped. But I had a day job with long commuting times. Ahead lay months of intermittent work on weekends and summer evenings. Exposed to water, rakìock wood does not last very long. I had to keep my kubbestol and dugout projects draped in unattractive tarps.

In the meantime, the enormous root ball loaded with soil and topsoil became an immediate jungle gym for our young son Jason and his friends. Summer rains rinsed off the sediment held together by the roots, producing an even better jungle gym.

Toward the end of the 1980 summer, several expert chain sawyers led by neighbor Mark Switzer kindly made two clean cuts, separating the future dugout canoe log from the butt end and its roots. Then, using ropes and blocks, they pulled the butt end away from the canoe log. The third chainsaw cut then separated the kubbestol segment from the remainder of the root mass. A fourth and final cut separated a 4-inch thick cookie (a round) below the kubbestol log. This hollow round still resides in our basement, waiting for some inspiration. A coffee table top, maybe? The rest of the root mass has long returned to soil.

CHAPTER 5

From Our Rakìock,
A Jumbo Kubbestol

In Norwegian, *kubbestol* means log chair, properly pronounced with the (accented) u as shortened oo as in cool, with the o as in an unaccented oh. These traditional, one-piece (plus seat), throne-like chairs are most commonly associated with Norway, particularly in the Telemark area, though they also occur in Sweden and Denmark. I first encountered them visiting Norway with my Norwegian-American wife, Randi. Some of her relatives had old kubbestols near their fireplaces. At the time, making my own never occurred to me.

Although kubbestols became popular in Norway around 1750, their origin is lost to history. The oldest surviving one dates from the Viking Age (793–1066 CE) and is displayed in Stockholm's Nordic Museum. A 5th-century kubbestol was found by archaeologists in Lower Saxony in Germany. The pagans of northern Europe were, plausibly, inspired by reports of Roman marble thrones.

Traditional kubbestols sport chiseled designs on their outsides; many are also painted. Most kubbestols have a faintly feminine, hourglass profile, with their "waists" narrowed below the seat level. A raised belt commonly encircles the waist. Vertical rills are carved in some, suggestive of skirts. This somewhat female form adds the grace lacking in stout cylindrical logs.

Some well-made or old kubbestols may have been revered as antiques and passed on through the generations. If kept inside, they would last a century or more. Some may have acquired superstitious significance. One kubbestol in the Norwegian Folk Museum has numerous milk teeth pounded into the frontal seat margin, perhaps to ward off tooth pains.

A kubbestol can be made from most any kind of tree of sufficient diameter of, say, at least 18 inches. Generally the chosen log is already hollow. If required, an existing cavity can easily be enlarged. Kubbestol backs are generally smoothly curved, resembling giant vertical shoe horns. Some larger carvings have backs curving around and horizontally notched on both sides as arm rests.

The wooden seat is either embedded in the cavity or riding on top, protruding over the lip. Some kubbestols : have removable seats—creating a great place to store and hide stuff. And they can be tipped back enough to hide things under the seat. Some also have small openings—perhaps heart-shaped— cut into their back rests. I have never seen a one-piece kubbestol, made from a solid log and hollowed out from the bottom up but leaving a seat. If one exists, I would love to see it.

Many old Norwegian kubbestols were made from birch. However pine, Siberian larch and Norway spruce logs have been used and are used today. The 5th-century kubbestol found in Lower Saxony was made from alder. Kubbestols, some elegantly ornate, are still made today—many from basswood (American linden) harvested in US forests. Unfinished and undecorated modern kubbestols can be purchased for a few hundred dollars, but the most elegantly carved ones run into the thousands.

The unfortunate resemblance of kubbestol seats to modern toilets can be masked with seat cushions, which are also easier on modern posteriors.

It would seem strange if Appalachian mountain homes did not have some versions, likely primitive and undecorated, of log chairs similar in form to simple kubbestols. Primitive log chairs, some just repurposed stumps, are even easier to make with modern chainsaws. But left outdoors, they seldom survive more than a year or two. Log chairs made from rakìock must be left dry indoors.

CARVING A TWENTIETH-CENTURY
RAKÌOCK KUBBESTOL

I joined the centuries of builders in 1980. The first task was prying the bark from both the dugout and kubbestol logs with a heavy duty, flat chisel. The diameter of the lower dugout log with bark removed was 31 inches. I discovered then that the tree was hollow with a 6-inch-diameter cavity at that lower end. I would have to deal with that for the dugout later. But for a kubbestol, this hollow was good news indeed.

Making the kubbestol indoors was not an option, so I made sure the would-be kubbestol always remained standing or lying on pieces of wood to protect it from the moist, critter-infested soil. And covered by a tarp when not being worked.

The next step was to enlarge the central cavity to 16 inches diameter and then to excavate the future seating space with lots of chiseling. I saved the larger chisel chips as "biscuit wood" to start fireplace fires. At some point, it dawned on me that there was enough room and design space to carve two arm-and-elbow rests into the interior. I doubt if any Scandinavian kubbestol ever had interior armrests, although a few larger ones have horizontal forearm rests notched into the back where it curves around and downward towards the front.

The reason for enlarging the seat cavity was to make this very heavy thing a bit lighter. rakìock density is 28 pounds per cubic foot when the wood is dry, which this log wasn't. Light for a hardwood, but it adds up. The final seat cavity I filled with a 2-inch-thick disk of rakìock wood sawn along the grain for strength. With some trial and error, I trimmed the disk to the right shape. I could simply pound it in place; the seat has not loosened in forty years.

Fearing I might ruin the project by sawing too far with my chainsaw, I drilled lines of holes with an electric drill, then used a mallet and wedge to break out the slab of wood in the front. I had planned after rough chiseling the kubbestol interior to decorate inside surfaces with uniformly random light chisel marks. However, by then the project had been outside for months, allowing shoe-string fungi to trace their lacy designs into the sapwood. These patterns, called spalting, do not substantially degrade wood quality. Spalting is sought after by modern wood workers, particularly for turning. Adding chisel marks would be decoration overkill, so I decided to sand the entire interior smooth. Rakìock wood is readily sanded, but the interior surface area was many square feet. The sanding took a long time. I did use an electric sander, but the corners and arm rests demanded hand-sanding.

The resulting combination of spalted and unspalted sapwood, and the contrast of pale sapwood with brown heartwood makes the kubbestol interior a work of art. At least its maker thinks so.

As for the large exterior of the kubbestol, I considered but did not cut a waist or carve designs. Nor did I use any paint. This was a clear departure from traditional kubbestols, whether made in 1800 or 2000. One might suspect my approach just avoids all that additional labor, and maybe reveals my lack of carving skills. Partly true. But I have carved many other wooden objects like ladles, bowls, trenchers, walking sticks, and in later years three more smaller kubbestols, and never decorated them.

My philosophy is to let the wood, plus or minus fungi or beetles, dominate the designs and colors. I would no more carve designs into or paint a kubbestol than get tattooed. Some smaller rakìock logs are beautifully rippled on the outside, just below the bark. Chiseling the outside with designs would destroy these ripples. But maybe someday I will cut a simple traditional belted waist around the base — no great carving skills are required—just to show off many of the annual rings, the story of the tree's life history in old age.

Once the kubbestol was sanded, I treated the chair repeatedly with linseed oil. My mother, an expert seamstress, later added a custom-made, reddish-brown pillow about 3 inches thick. It covers the entire large seating area. More comfortable, and insurance against occasional tasteless comparisons of my prize kubbestol seat to a giant toilet.

The final product stands 3.5 feet tall and, at its level, curvaceous top, is up to 37 inches across. The 23-inch wide seating area has accommodated even the widest visiting butt. So far. If the many visitors who have seen and perhaps sat in my rakiock kubbestol found it too stout and clumsy looking, none ever said so.

My wife was not thrilled about such a massive chunk of wood furniture dominating our living room. In fact, she secretly hoped it would not fit through our front door. One day in October, 1981, some strong teenagers muscled it barely through the front door and up some stairs into the living room, where it resides to this day.

Installed in our living room, this giant chunk of wood began to dry. Moisture inside the wood began to equilibrate with the average interior humidity. Southern Maryland has a humid climate, but when outside air is pulled inside houses and heated in winter, relative humidity declines. The wood in the kubbestol adapts to its average environment. Moisture escapes wood most readily along its pores, from both the top and bottom of the chair and from the seat. However, by slathering linseed oil on the wood, I slowed this escape.

As wood dries, it also hardens. Wood carvers scavenging fresh windfalls are advised to rough out their projects while the wood is still green. That also reduces later checking, cracking and warping. Even when repurposing wind-thrown trees just for the fireplace, it's easier to split green wood.

After some weeks, maybe a few months, I noticed a number of very narrow cracks inside the back above the seat. I was sure these small fissures were no threat to the mechanical integrity of my kubbestol. However, they were unattractive, so I duly patched them with appropriately tinted wood filler.

In time, two other cracks began to appear in the base, extending down from the seat at the corner where it joins the back. As these fissures widened, I filled the growing gaps at intervals with slivers of wood pounded and glued in place. As that was happening, the many narrow fissures in the back began to close, squeezing out the wood filler I had applied earlier. My attempt to arrest the cracking with tightened belts was futile. The drying osmotic forces are far too strong. It took at least two years for this kubbestol to stabilize.

A large mass of wood drying and cracking, once brought in from nature, is worlds away from store-bought furniture, made of thin saw boards or veneer, machined high-tech from selected, defect-free, kiln-dried wood of uniform texture, milled and chemically treated. I contemplate this while relaxing in my kubbestol, as if joined to the living tree.

The initially contrasting, pale slivers of wood filling the cracks gradually blended into the yellowish brown of the chair, with the heartwood retaining its original darker reddish brown. The top, or rim, has acquired an antique shiny dark brown look from hundreds of human hands over four decades of use. It is a potpourri of human and plant biochemicals.

All that remains is more visitors to admire and in many cases sit in it. Very comfortable indeed. For those whose legs are too short below the knees, I offer a tripod black locust stool as a footrest. The mammoth kubbestol is a kind of throne, ruling our living room as the living tree once ruled our yard. May it have a long life, long after we—as Hamlet put it—shuffle off our mortal coils. Barring house destruction the only threat is from some future owner putting it outdoors as yard furniture, for it would rot away in no time. Lacking any insurance against such an evil deed, I used a fat felt-tip marker to write a curse inside the base. Of course, who would go to the trouble to turn over this massive chair to read the curse?

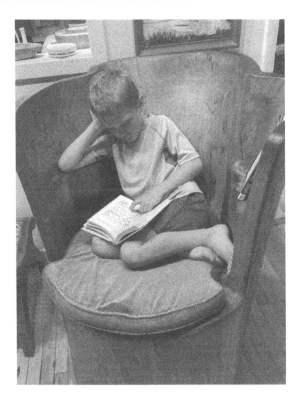

CHAPTER 6

A World of Rakìock Creations

In spreadsheets with wood characteristics, no species scores a five in every column. rakìock wood, for example, is relatively soft, weak, and lacking in figure. But every kind of wood has its own qualities and potential. The cylindrical form of branches and smaller tree trunks of many species—particularly hollow ones—can be fashioned into all kinds of items, including containers that are both decorative and useful. Elongated blanks can become trenchers—miniature dugouts. If the entire cylindrical form is retained, the blank, if hollowed, can become a cup, a wooden beer stein, a beaker, a container for kitchen utensils or other knick knacks, or a humble waste paper can. I carved one hollowed hickory log into a 3.25-foot-high, 15-inch diameter hiking stick or umbrella repository (or alternatively the log might have become a junior kubbestol). A smaller diameter hollow red oak log, too small for a kubbestol, became a long-necked kitchen stool (or a stand for coffee mugs of couch potatoes). Its seat is a round plate of sassafras.

With no lathe, I depend on large wooden cylinders. Turners have no such dependence; they shape round, sometimes eggshell-thin containers from formless blocks of solid wood. I have so much wood and so many tools in my small basement that there's no room for a lathe. Maybe in my next life. But I do use other power tools: angle grinders, belt sanders, a Dremel tool (for cutting, sanding, engraving etc.) and other high-speed cutting and grinding tools. I have made trays and platters from boards rough-sawn by the Amish from downed logs taken there by a neighbor with a pickup. I do own a draw knife but have not yet built or bought a shaving horse: Where to fit that in a tiny basement?

Because none of my smaller creations, variously of hickory, oak, sassafras, black locust, dogwood, cherry and black walnut—all foraged locally outdoors—are rakìock, their stories remain outside this book.

In the years that followed my jumbo kubbestol, I carved three others, one for a young grandson. Two of those are of rakìock; one began as a giant piece of sycamore driftwood washed ashore along the Chesapeake Bay. Those three chairs are all of more typical size, and all but one remain undecorated by human hands. In a concession to tradition, I cut a waist into the base of the sycamore chair. It brought out the spalting! The rakìock kubbestol made for our grandson sports his initials chiseled in gothic-style letters. This chair and the sycamore one have seats covering the central cavity, i.e., not embedded in it. The seat plates on the sycamore chair and embedded in the second rakìock chair are repurposed, air-dried, foraged cherry heartwood. Red and hard, the cherry seats will likely never crack.

The cherry wood came from a wind-thrown black cherry, and the slab used for the embedded seat plate had a wide irregularly shaped cavity, which I patched with slivers of rich brown sassafras. The second rakìock kubbestol had its brown heartwood honey-combed with large beetle galleries. Instead of patching up and thus camouflaging these cavities with tinted wood filler, I filled them with conspicuous white filler so as to let these long dead beetles help decorate my kubbestol. Nitpickers—if any—might compare this bright natural decoration to splattered white paint or to a Holstein cow. I would remind them that when and if humans disappear from this planet, members of the vast beetle family will still be here munching their way through wood.

In making defects decorative, I took a page from modern turners who fill natural cavities with granulated turquoise paste.

I began to make three more rakìock kubbestols but ran out of interest and energy. These semi-completed ones, lacking seats, still languish, protected under our house from the elements. They somewhat resemble those unfinished stone statues on Easter Island. I am happy to give them to any enterprising kubbestol fan to complete what I began.

JUMBO RAKÌOCK ROUND TABLE

A disk-shaped round or "cookie" sawed off from a log of any diameter is the source of the simplest useful objects. (It is easiest to make the two parallel cuts perpendicular to the log.) Smaller disks, easily made with hand saws, become tree ornaments— maybe painted, or with something interesting such as a shell, arrowhead, or fossil shark tooth—glued on their surfaces. Larger disks become coasters or serving trays.

Still larger cookies can be made into coffee table tops or night stands wide enough to hold a table lamp; an elliptical cookie makes a better coffee table. The larger the diameter, however, the more challenging to make the two completely parallel cuts free hand with only a hand or a chain saw. It's still more challenging to maintain a constant thickness when the log is twice sawn diagonally by eye. Saw cuts not completely parallel produce rounds very laborious to level with angle grinders and belt sanders.

Cookies more than about 6 inches in diameter generally crack upon drying, especially indoors, though racking depends on the wood and its moisture content. (Maybe not if you live in the tropics and keep your screens open all year.) Such cracks can be inches wide for disk diameters of a foot or two. Cracking never causes cookies to fall apart, so the issue is only appearance. You can patch up the crack with similar or contrasting wood, or live with it. If you fill cracks with wood filler and/or wedges of glued wood, larger rounds will expand in the humid summer and contract during the course of dry indoor winter conditions. The central crack will close during the summer and widen somewhat by late winter.

After our community had its largest tree unnecessarily cut down about 1986 as I described earlier, the massive trunk was left on the grassy ground. Time went by. The dead rakìock began to rot, starting with the bark. Some months after it went down I went to take a look. Maybe some of the wood was worth repurposing. In diameter it was far too big for a kubbestol. Someone had already tried to saw the trunk into sections; however, no one-man chainsaws were available—if indeed any exist—to slice through a 6.5-foot diameter log (including bark). I noticed that the main butt log had been basically butchered with saw cuts from several sides, not matching and not parallel and not all going through the log. I never learned who had worked on this log and what they planned to do with the log sections. Firewood? Lots and lots of wood but of low quality.

I considered hiring a tree service to salvage three large cookies from the least butchered parts of what remained. Such services must have longer chainsaws and more skilled sawyers. I certainly lacked the skill and a long-bladed chainsaw. The hired sawyers deposited three giant cookies, slabs ranging from 4 to 10 inches thick, in my driveway. Their saw cuts still did not line up well and were not parallel. But I did my best to even out these mistakes with angle-grinders, belt sanders and other tools, running an extension cord from our house to the driveway.

The plan was to donate two cookies to the Battle Creek Nature Center. Rounds once sanded and finished are often displayed in natural history museums for nature education—pins with years may be pinned on various annual rings to mark local or other milestones. The nature center was happy to receive the two jumbo cookies; one was later either traded, donated or sold to another museum. I have no idea of its fate. The slab they kept was used for many years as a base for stuffed wild animals. The base, a table in one corner of their small auditorium, was seen by many hundreds of visitors, especially schoolchildren. Years later, the staff reorganized the exhibits. The raklock round had some cracks and other defects on one side, so staff cut off that half and displayed the better half lying on the exhibit room floor, where it resides today. The best and most uniform of the three slabs I finished and offered to our community's small museum, the original rustic summer home of the founder and developer. The slab was exhibited there, leaning against a wall. However, the volunteers who maintained this museum—also rented by members for special occasions such as dinners—decided there was just not enough room so they returned my gift. Would that be called "Indian Receiving"?

I said thank you, our family will use it as a jumbo table top in a small screened porch. So one of our sons built a temporary stand from spare dimension lumber and—with the help of several muscular teen friends—hauled it through two doors, vertically, strapped in by ropes, into the porch. This giant round became and remains a perfect party round table. Up to ten guests can pull their chairs up to it. The slab still rests on the temporary but well-built base and has not been moved. It has enabled many dinners and hours of chatting during long warm-month evenings. We eat and talk by the light of candles or, more recently, the light of a table-top fire pit fueled by alcohol. Sometimes we also have special optics from summer lightning.

Exposed to rather uniformly high-humidity outdoor air, the 1-inch-wide filled crack moves little from summer to winter. The slab had some modest crevices near the center that I patched with wood slivers early. The table surface has long had numerous tiny cracks, none of which have ever widened. Perhaps moisture moves into and out though those fine cracks.

Screens and roof overhangs protect against soakings, but during windy rain or snow storms tiny droplets of moisture get blown through the screen mesh to moisten the table. These small additions of moisture and the average high outdoor humidity may explain why the sapwood rim closest to the outdoors has been locally infested with gnat-sized powder post beetles. I recently poured wood hardener on those places to harden any damaged wood below the table surface. Other than that, I give the round table an annual light sanding and spray on polyurethane. The many pinprick holes left by hatched beetles are part of the decoration and story, but I won't tolerate any part of my prize cookie reduced to powder.

A human generation has passed since wind and gravity felled this massive rakìock, and most of today's residents don't remember the tree. Its memory lives on in this table, likely the largest round far and wide. And young guests enjoy the challenge of counting the 155 annual rings.

CHAPTER 7

Under Wraps at Battle Creek

Hiking in local woods, I often took a trail through a swale dominated by majestic, centenarian raklocks. Foresters would call these trees mature or even over-mature and recommend the time was right for their "harvest." However, these trees are today safe to age in place, as part of the American Chestnut Land Trust, which is largely managed as a nature park.

Some foresters I have hiked with studied the ground for signs of oak regeneration. They correctly call the raklocks a pioneer species, like Virginia pines growing up on abandoned fields in other local soils. However, from my half century of observation, raklock is both pioneer and climax forest. More about that later.

In my walks through this even-aged but fairly old forest, I pictured that it was used for either grazing or tilling land in the 19th century and already well forested by 1938, the year Calvert County was first photographed from the air. Pine-covered areas look conspicuously dark from above, indicating fields cleared prior to 1938. But no pines were evident where my raklock swale is today.

However, I always noticed one raklock, about 100 feet from the trail, larger and likely older than all the others. Leaning, hollow and not worth cutting for saw timber, it was left to provide shade for people and animals. This picturesque giant must have been about 4.5 feet in DBH (Diameter at Breast Height) but with a much wider flaring bottom.

Year upon year, storm upon storm, this old specimen stood up to the elements. Then one day I noticed he or she had fallen, broken along the ground, leaving its flaring, fluted base—free of actual roots—up high. I thought again of Shelley's poem about the shattered visage of Ozymandias, king of kings.

One more crazy project with a giant rakìock log? I already had a kubbestol and a dugout, as you will see, and this log was too irregular and hollow for either. I would figure that out later.

So I first asked a neighbor with a 30-inch chainsaw and more sawing skill to cut the butt end of the log into three large chunks, starting with the lowest, largest and most complex, and two other massive ones. There was no power outlet in the forest and no space in our small parking area, so I persuaded the head of the Battle Creek Cypress Swamp Nature Center (and other Calvert County nature parks) to let me work on these jumbo pieces near the nature center. Dwight Williams reluctantly agreed, with the understanding I would explain the project to park visitors walking by: what I knew about trees (especially raklock); wood-decay processes; wood-boring beetles; the tools I was using, etc. In the following years, Dwight probably regretted his decision.

Meanwhile, the dead giant had been lying on the forest floor for months. The bark had become loose, so I split much of it off.

I hired a local contractor, Morgan Russell, to send his front-end loader into the woods, and with the help of a large truck, bring the three logs to the Battle Creek Nature Center. The two smaller ones were placed inside the roofed, open-air pavilion, not far from the nature center. Inside the concrete-floored picnic space were two sockets. The largest piece would not fit under the pavilion roof and so was set just outside, still within easy reach of my extension cord. It stood—like a short, fat, dead tree—with the wide flaring bottom up, resting on pressure treated four-by-fours. To keep the rain off, I draped a large brown tarp awkwardly over the project.

Hauling chisels, hatchet, saw and a Dremel tool to the site the next day, I pulled off the tarp and set to work. Splitting off the remaining bark was the first and easiest chore. I knew there would now be not hours or days, but weeks or months of inter-mittent work just "cleaning up" the interestingly irregular top rim and the outsides. By that I mean filling cracks and defects with wood filler, sawing off and then smoothing awkward, fragile or splintery protuberances, and filling punky wood with wood hardener, aka stabilizer. I squeezed wood filler into small cracks and glued wood into the wider ones.

But what exactly was this supposed to become?

The large interior hollow actually contained one or more coaxial cylinders of dark brown, very hard and strong heartwood. These structures were attached to the rest of the stump at a few places. Leaving them would make for a really cool sculpture, but there would be no way to reach down to clean up them or the interior walls of the main stump. I still don't understand the biochemical processes that created this horn-like wood, when the heartwood originally filling this cavity had long since rotted and fallen away. This was not your typical old hollow tree.

I decided to remove these cylindrical shells and open up the interior. The first step was to widen the entrance. While still in the standing tree, this gate to the interior was wide enough for the world's fattest groundhog, but not for a bear or an adult human.

Using saw, ax and hatchet, I created a portal 18 inches wide. Once those mysterious, horny cylinders were removed, I could sit in a folding chair inside the piece so as to widen and clean up the inside walls. The tree trunk, and hence the interior space, were somewhat elliptical in cross section. My cozy hiding place measured 30 by 38 inches at the base. Halfway up, the interior widened to about 4 feet.

Whereas the outside surface of this upside-down stump was uniformly light brown and fairly smooth, the inside was a mix of dark heartwood and lighter sapwood, both irregular and honeycombed with beetle galleries. Quarter-inch to half-inch wide holes incriminated those large shiny black wood boring beetles, but I never found even a dead one. All this munching evidently happened long ago.

Filling the many beetle galleries exposed in the inside walls was an exhausting option. I decided instead to preserve most of the beetles' contributions. It would become their part of the sculpture and its story. The sharp edges of the gallery holes were smoothed with the help of Dremel cutting and sanding bits.

With a day job in DC and having to drive to Battle Creek Nature Center, I was there only intermittently. Most park visitors only saw a tarp covered mystery. Many likely considered it an eyesore. The chief naturalist/park manager let me know he wanted the project finished.

Weeks turned into months and months into years. Storms blew the tarp off and collected water in the depressed center. The first tarp eventually tore and had to be replaced. Meanwhile I had many interesting encounters with some of the hikers who passed by. Many, of course, just walked by, perhaps wanting to avoid talking to a bearded weirdo, or perhaps too shy or incurious. While I worked sitting inside the project I could not see anyone outside. Wearing ear muffs and working with my Dremel, I couldn't hear voices or footsteps approaching. So I was frequently startled by strange human faces peering down at me from above the rim.

THE TROLL PULPIT

From many encounters with park visitors three stand out—

While sanding the outside surface of the project one day, I noticed two men and a woman hiking together in my direction. I did not recognize any of them. When this trio approached, I stopped my work and began explaining what this was about. The woman, who identified herself as a Unitarian minister from upstate New York, offered to buy the piece as her pulpit. She would arrange for a truck. That was certainly a new and intriguing use. I said it wasn't yet finished but I would be interested. However, as they prepared to continue their hike, I pointed out that I was not really religious. So using my project as a pulpit would be ironic. She smiled. "I have plenty of your kind in my congregation," she said. We failed to exchange names and addresses and I never heard from her again. As a result of this encounter, I began to call my project the Troll Pulpit.

Battle Creek hosts many visits from school children, mostly arranged by the Calvert County Chespax Program. Each year, every age group is treated to one outdoor environmental education trip, generally within the county. Schools from outside the county also come to Battle Creek to visit the nature center and, on the boardwalk, admire the so-called trees with the knees, the stands of cypress. One field trip route takes the class past the picnic pavilion and my Troll Pulpit.

I'm not sure if this blond kindergarten-aged girl was part of a family or part of a class. But upon looking over the Troll Pulpit she appeared displeased. She glared up at me and asked, "Why didn't you leave it alone?"—thinking I had cut this tree down. Clearly, this girl was destined to become a tree hugger and environmentalist. I explained that the tree was taken down by gravity and wind and that most of the wood had already nourished beetles, fungi, woodpeckers and whatnot. I had only taken a part of the wood and was using it to teach people about trees. She seemed satisfied with my answer. She is probably a park naturalist or arborist by now.

Some months later Battle Creek hosted a field trip from Baltimore. A young school-age girl studied the Troll Pulpit but said nothing. Noting her interest, I asked what she thought of it. Not missing a beat she said "I'd be ashamed to have that in my house." I figure she is now an interior decorator. Her opinion seemed to foretell my later failure to sell my finished masterpiece. Of course I was not honestly suggesting it for anyone's living room, but I imagined it installed in an airport or lobby of a large modern building to offset all the metal, plastic and concrete.

By 2009, the Troll Pulpit was finished, but before any marketing I had to measure it. The base (about where the diameter at breast height would have been measured on the standing tree) was elliptical, with axes of 54 inches maximum and 43 inches minimum. Using a formula, I first added 5 inches to these numbers to account for the original bark and calculated the original circumference of the standing tree at 167 inches, about 14 feet. Had the trunk been circular, the original diameter would have been 54 inches.

At the top of the piece, the diameter was 90 inches maximum and 60 inches minimum. The height varied around the perimeter, but ranged from 48 inches in the lowest saddle to 76 inches for the highest spire. I never weighed the irregular shape; how heavy was it? I approximated it with a rough estimation for a hollow cylinder with wood density 28 pounds per cubic feet. This yielded 1,000 pounds, but my gut tells me it's not quite that heavy.

Time marched on and eventually the park director retired—with the mystery eyesore still in place. The new director, Karyn Molines, had not been there long when she gave me a polite ultimatum. Either find another home for the three wood chunks or just let them remain outside uncovered and return to nature.

I had no bargaining power but first offered to let her staff pull one of the other two pieces out of the pavilion and let kids climb on it until it crumbled with age and rot. My masterpiece, the Troll Pulpit, might look really cool inside the nature center, I implored. It was part of nature and told the public about big old hollow trees. In olden days, when black bears still roamed around here, would not this have been the ideal place to hibernate?

The director was less than enthusiastic, but she discussed the matter with her staff of naturalists. She worried this would become a kind of Trojan horse, with all kinds of vermin escaping from beetle holes into the warm nature center. The staff was divided, but assistant director Andy Brown—on my side—won over the others.

My beetle gallery-decorated masterpiece entered the nature center in 2013. The top of this upside down stump was, however, too wide to fit through the sliding double door. My ally had to saw off one corner. Then the machos from Calvert County Buildings and Grounds, using some giant dolly, rolled the half-ton Troll Pulpit inside. The flat surface created by sawing off the corner became the perfect place to affix an explanatory panel.

The Nature Center renamed and repurposed the piece as a reading nook. A pillow on the floor inside the stump lets visiting kids get in and sit or curl up with or without a book. Since then, many visitors have admired (I want to believe) this wooden giant, and photographed their kids or grandkids inside. Some visitors assumed this was from a bald cypress tree; after all this was the Battle Creek Cypress Swamp. So the sign was revised to identify this as a rakiock.

I don't worry about the reading nook. However, should a future nature center director place it outside, that would be its demise. Short of that, a heavy teen climbing on the rim could break it off. A fire could consume the nature center. There has been one fire, in November 2022, but only the basement was destroyed.

CHAPTER 8

The Dugout Canoe
in History and Prehistory

A hollowed-out tree trunk was likely humankind's first boat, invented independently wherever trees grew to sufficient size and relatively flat water bodies—large rivers, lakes and estuaries—were close by. The oldest surviving dugout—found buried in a Dutch peat bog—was carved about 9,200 years ago. Only 10 feet long and 17 inches in beam, the "Pesse" canoe nears the smallest practical limit for even a single canoeist.

Dugouts several thousand years old have been found in Africa, Asia and the Americas. Since larger canoes were less common and are less likely preserved, the upper size limit for ancient dugouts is unknown. However, a Bronze Age dugout 35 feet long and 2 feet deep was found in Britain; one of similar length was recovered from a North Carolina lake.

The 3,000 year old, 14.5-foot-long dugout retrieved from the bottom of Lake Mendota along the shores of the University of Wisconsin in 2022 might well have been made by Proto-Algonquian-speaking ancestors of those English colonists encountered in the Chesapeake Bay. The ancient Wisconsin canoe was made of oak, but as Algonquian peoples migrated east towards the Atlantic and then down the coast, they likely brought along their dugout-making traditions. Linguists estimate that around 1000 BCE the ancestors of Chesapeake-region Algonquians were living in what today is New England and northern New York state. They were speaking Proto-Eastern Algonquian, an extinct language partly reconstructed by modern linguists. This is also the time-depth of steatite (soapstone) bowls, used by natives on the Mid-Atlantic Coastal Plain from around 1500 to 500 BCE. Transporting such heavy objects long distances east from Piedmont quarries likely required dugout canoes, so the discovery of such bowls in archaeological sites in the Chesapeake region is indirect evidence for ancient dugouts.

The Stone Age Pesse dugout was carved from a Scotch pine log. Many other species have been used as well. In North America this includes white pine (New England), bald cypress (southeastern Coastal Plain, US), white oak (Great Lakes), cedar (Pacific Northwest) and cottonwood (the 'pirogues' of the

Mississippi-Missouri). In the Chesapeake Bay region, loblolly (yellow pine) was used particularly on the Eastern Shore, and rakìock (tulip or yellow poplar) in Southern Maryland and elsewhere in its natural range. A Colonial chestnut dugout was found on the York River. To what extent bald cypress (e.g., from the Battle Creek and Nanticoke River swamps) were also used remains speculative. Wood from conifers, cottonwood and rakìock is around half as dense as that of oak or hickory. The ponderous 550-pound dugout I would carve would have tipped the scales at half a ton, had it been carved from such species.

Some past canoe builders made their dugouts beamier by filling them with water, heating the water with hot rocks, and then, when the sides were softened, prying them apart with poles or (later) boards wedged between the gunwales as spreaders.

Bark canoes were also used in the Chesapeake tidewater. Birch bark canoes might have been obtained by trade or capture from raiders coming south along the Susquehanna. John Smith wrote that canoes were also made of "the barkes of trees, sewed with barke and well luted with gumme." The Nanticoke Indians traveled to the mouth of the Susquehanna in the spring and stay "till the Barque will peel soe they can make Canooes." However, there is no record of any large canoes made that way. The type of bark was not reported, but rakìock bark can be stripped off in June when the sap flows, and was also used as roofing for witchotts.

The native burn-and-scrape method of making dugouts was observed by Hariot in 1585 and again by John Smith twenty-three years later. As Thomas Hariot described in detail, they made a fire around the base, using wet mud or moss to keep the fire from burning too high up the trunk.

After the tree fell, they burned off the top. Referring to the dugouts, Smith wrote that "These they make of one tree, by burning and scratching away the coles with stones and shells till they have made it in the form of a Trough."

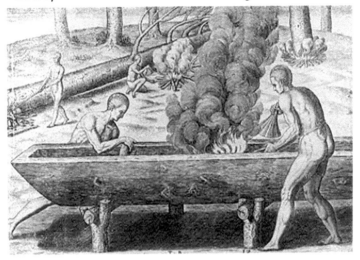

Most Chesapeake area native dugouts were likely in the 15-foot range in length, but early explorers, notably John Smith, reported some more than 20, and up to 40 to 50 feet long, and up to an elne (ca. 3 feet 9 inches) deep. Assuming a length-beam ratio around 7 to 8, these larger dugouts required debarked tree diameters from 3 to 6 feet. Trees of that size surely existed: One explorer noted that three men could scarcely get their arms around some trunks. These big canoes could carry from 20 to 40 men, along with their gear. We can only imagine the dugout sizes achievable by natives, had even larger trees existed. Laying eyes on a large European ship like the *Ark*, some Indians, assuming it had to be some kind of dugout, wondered where trees of such enormous size were to be found.

English explorers and early colonists were impressed by the speed with which the natives propelled their primitive craft. Instead of rowing with oars as Europeans did in calm weather, the natives used "paddles and sticks with which they will row faster than our Barges." In 1605 another traveler had noted: "Our boat well manned with 14, yet they row faster with 3 Ores in their Canowes then we with 8."

For similar hull shapes, the hull speeds of boats increase as the cubic root of length. Thus, 45-foot and 30-foot long dugouts would be respectively 44 percent and 26 percent faster than the 15-foot rakìock I would build. It may be that the particular canoes admired by the English were longer than the so-called English barges with their beamier hulls reducing hull speeds.

When English colonization began—a generation earlier in the southern Chesapeake in Virginia than in the Maryland shores—the natives began to fashion dugouts with iron tools. At first the English derided native dugouts as "hog troughs" but soon began to make their own, or hire natives to carve them. As M.V. Brewington explained in 1963, the Maryland colonists brought over skilled boatwrights, while the unprepared Jamestown colonists came to depend on dugouts early, given that little skill was required.

Colonists soon began to improve and expand on the native dugout. The first multi-log dugouts had already appeared by 1687, beginning a uniquely Chesapeake boat-building tradition. By 1870, the dugouts, now evolved into sailboats, had from three logs to as many as seven. The evolution away from single logs was driven by the need for larger craft but also to the depletion of old-growth forests with large trees.

In a different adaptation, two single-log canoes were attached to form a catamaran, with tobacco hogsheads arranged athwart ships in two rows, resting on planks connecting the two canoes.

However, single-log dugouts of large size continued to be made wherever big trees still grew. In 1799, a disgruntled Daniel Boone hewed out a 60-foot raklock dugout, into which he piled his family and possessions, leaving Kentucky down the Ohio into Spanish territory. Of course he just had to steer while drifting downstream.

Arguably the most famous historical one-log dugout in Chesapeake tidewaters was the *Methodist*, hewn from a famous large loblolly pine growing in Somerset County. In 1806, a man named Hance Crosswell purchased this loblolly pine for $10 and proceeded to ax it down in a few hours. If any neighbors were saddened, it wasn't recorded. The larger half of the trunk log was hewn into a dugout, the hull shaped into a standard, streamlined form. Some 20 to 30 feet long, with a 5-foot beam, the boat was outfitted with sails. In about 1831, this unique single-log sailboat was purchased and rechristened by the Methodist preacher Joshua Thomas. For decades thereafter, the "Parson of the Islands" made his rounds in the boat. Thomas died in 1853, but his boat lasted until 1907.

Today, a few multi-log dugouts still sail on Chesapeake tidewaters. Some prehistoric single-log canoes have been found buried in oxygen-poor mud. A few of those have been carefully excavated and chemically preserved, including one at the Maryland Archaeological Conservation Laboratory at Jefferson Patterson Park in Calvert County. The largest assemblage of buried or semi-buried native dugouts—more than 30—were discovered starting 1985 in Phelps Lake in eastern North Carolina. One is 30 feet long, but due to poor condition was left in the mud.

A few early historic ones survive, like the one exhibited in the Valentine, a museum in Richmond, Virginia. Some replicas have been made. The simple native form, replicated in 1980 from a raklock log with the help of chainsaw, is on exhibit at the Calvert Marine Museum. Next to it rests a 19th century one-log punt fashioned with iron tools of that time. Both craft were once paddled for very short distances in calm water.

The native method of burn-and-scrape has, and is, being demonstrated, notably at Jamestown Settlement, a museum of 17th century Virginia history, and in Maryland at Historic St. Mary's City and Jefferson Patterson Park and Museum. However, most of the dugouts produced have never seen the water. A large dugout made years ago at Jamestown Settlement was launched but not paddled far, to my knowledge. No modern replicas of single log canoes had been repeatedly paddled for miles and miles. *Rakiock* would change that.

In modern times, one native-type dugout was reportedly built by burn-and-scrape and actually paddled. The young man who accomplished this, Russell Reed, was a historical interpreter at Jamestown Settlement and a native of Mathews County, Virginia on the Chesapeake's western shore.

CHAPTER 9

Carving *Rakìock*, 1981-1983

Already in the 1970s, I had the dugout bee in my bonnet. I live among many trees, some with suitably long straight trunks and base diameters of several feet. However, all those trees were standing. There were some windfall logs here and there, and I began hollowing one out. But it was marginal in size—only sufficient to match that 10,000 year old Pesse dugout found buried in a Holland peat bog. So I lost interest. I suggested to the director of the Calvert Marine Museum—where Chesapeake boats and maritime history are major themes— that they should have a dugout. Maybe, I thought, I could make one and donate it. Then the storm deposited a proper log, and of the right species, in my backyard. The kubbestol would delay my dugout project by only a few months.

I would have preferred working in our level parking area, but that was more than 100 feet away and, more importantly, more than 20 feet higher. Winching a one-ton log past the trees and around our house? Probably not. And to make it even harder, there were many trees in the way. So I decided to make the canoe where it fell, on a slope. The first steps were steps— spading steps into the hard clay soil on both sides of the log. It was then a compromise to use modern human-powered tools similar to those used by colonists—and by the Indians once they acquired English tools.

My rakìock log was, however, not a straight cylinder, but somewhat curved around two axes. The greatest curvature became the bottom, which was convex downwards. That was actually okay; the bottom of the hull would be a bit curved, thus giving the dugout a few degrees of deadrise. The other curvature was not so good. Whoever heard of a bent boat? I could cut enough wood from one side at each end, and on the other side in the middle. But cutting wood off the sides also meant reducing the beam. That would mean less space for the paddlers. So I compromised a bit, leaving a bit of curvature. I hoped this would not impact performance too much. If it did, I could always ax off more wood. Lacking an old-fashioned broad ax, I removed some side wood with a hatchet and a chisel—very carefully. I was working on soft sapwood.

Next I drew the planned gunwale outline on both sides of the log and marked lines across the top for the planned saw cuts. A friend and I then sawed cuts about every 6 inches down to the gunwales. rakìock wood is easy to saw. It was easy, fast and fun to whack off about two dozen blocks, starting at one end of the boat. Did I use a sledgehammer or a my-handmade wooden maul (a club-shaped piece of hickory)? I can't remember. But those blocks were the perfect size for the fireplace. Rakìock wood burns fast, so it's okay when you don't want the fire to last long.

After that it was weeks of intermittent chopping. I attached thin boards to the gunwales to protect them from being unintentionally chopped. After the dugout interior was roughed out, the finer cutting was done with an old-fashioned lipped adze. Still finer details were managed with planes and curved chisels and pounded with a one-piece mallet I had made from black locust.

Our ravine frequently echoed with the sounds of whacking and pounding, punctuated with old-fashioned seaman oaths.

Neighbors took notice. My wife was aghast at my language. Local kids would show up at times to inspect my progress. "Gee, Mr. Vogt, are you still working on your canoe," I would hear, accompanied by guffawing.

There remained one more production challenge. My log was hollow at the stern end. This roughly 5-inch diameter hole had to be filled. And my logistic challenge required moving the finished canoe—first up to the road, then once launched frequently out of the water onto a beach and onto the back of a pickup or flatbed truck. From one waterfront location to another. Natives and Colonists did not have such problems.

My solution was to chisel the cavity into a tapering somewhat conical shape. I took a small black locust log and shaped it to fit into the cavity, with the narrower end aft. Then I drilled a hole along the axis of the locust plug and inserted a robust eyebolt through the hole, with the eye protruding aft. Ropes or chains could be fastened to this bolt to haul it stern first. The stern eyebolt and creosote finish are of course concessions to modern times. The tapered locust plug—glued into place—could thus not be pulled out of the canoe.

Sitting or kneeling in the bottom of any canoe is not very comfortable. So I decided to carve a seat into the stern. As is not possible in typical modern canoes, I would be able to sit very close to the stern. The big eye bolt nut protruding into the stern would now be hidden from view. There would be some dry space under the seat to stow away stuff. Carving this seat was laborious and time-consuming, but could well have been done in olden days. I can't prove my stern seat, carved out of the same log, made my dugout unique.

Most of my work on the dugout was done in 1982. With the canoe basically finished by early 1983, I hoped to add some color or carvings on the outside. Of course, not as elaborate and skilled as what tribes in the northwestern United States did with their ocean-going cedar canoes. However, I could find no historical or archeological evidence that any of the native-built dugouts of the Chesapeake tidewaters had any kind of decoration. Very disappointing! My dugout and kubbestol projects were thus in stark contrast. For the latter, I rejected the traditional carved and painted decorations and opted for plain wood. For the dugout I had hoped to add decorations but rejected them because that would violate traditions.

I used to deprecate our Chesapeake natives for not decorating their dugouts. Those Pacific Northwest tribes were evidently superior. However, dugouts for our Indians were what buffalo horses were to the Comanches. Dugouts were necessary for locals to travel and fish. They could be stolen. More importantly, they were obvious evidence of nearby villages. Dugout decorations would have signaled "steal me" and "plunder available nearby." Enemies of tidewater Algonquian tribes came down the Susquehanna and attacked from water. You would want your boats to resemble driftwood logs. To hide them was even better, sometimes canoes would even be submerged with a load of cobbles. Decorating dugouts would be worse than leaving keys in your pickup.

Some native dugouts may have had keels attached to their bottoms. At least that was what James Michener wrote in *Chesapeake*. I wrote to him to find his source, but he deflected the question by suggesting I talk to some local archaeologists. Obviously he had made up the keels! That's a risk of turning to historical fiction as a reliable source for history.

Now that my dugout was finished, I wanted to preserve it as long as possible. Raklock wood does not last very long exposed to rain or lying on the ground. During the months it was unfinished, I kept a tarp over it when not actually working on it. Now that it was basically finished, I brushed the vessel with creosote on the sides and bottom. Yes, at that time creosote was still legal, its hazards yet unproven. The dugout insides were painted with a mixture of turpentine and linseed oil, presumably available to English colonists.

The burn-and-scrape method probably helped preserve native dugouts because charring seals the pores. Charcoal lasts forever. Maryland farmers using long-lasting black locust fence posts charred the post ends prior to embedding in the ground. During construction I did experiment with burn and scrape, even grilling some hotdogs with the fire. Based on that very limited trial, I very roughly estimated that two adults and an older boy could have made one by burn and scrape from a tree in about 10 days, while camped near the log.

An Indian-type dugout had been built in 1980 for the Calvert Marine Museum. It was similar in size, shape, and wood (raklock) and the charred inside made it appear authentically crafted by the scrape-and-burn method. In fact, the dugout had been cut out largely with a chainsaw, then charred artificially. This was an early project by the new museum Patuxent Small Craft Guild, which I soon joined. As a club member, I helped construct a 19th century dugout punt of the same size, also from rakìock but sharp-ended, streamlined, and at 200 pounds lighter—and, when launched briefly at Chesapeake Marine Museum—faster. But it also was much less stable.

My completed dugout was 15 feet 3 inches long, with a beam width of 2.5 feet. The length/width ratio (6:1) is a bit less than for many dugouts (7 or 8:1). According to Brewington (1963), the depth to beam width varied but was generally one-half to two-thirds. For my dugout, this translates to a depth of ca. 15 to 20 inches. My dugout was carved only to a depth of 8 to 9 inches below the gunwales, which range from 1.5 to 2.5 inches wide. Given the scarcity of recorded data, I erred on the side of leaving more wood in the bottom. Once adzed off, wood obviously can't be replaced.

The early Colonial-era dugout at the Valentine Museum in Richmond is somewhat comparable. It's 15 feet 6 inches long by 2 feet 1 inch at beam. Its 18-inch depth is twice that of mine, making it somewhat lighter. Experts think this canoe shape suggests some English influence, so it may have been made about 1650.

As I made my dugout, the thick round belly (up to 8 inches wood below the centerline) improved stability and space for a wider flat bottom. Judged from a photo, the rounded stern of the Valentine dugout looks wide enough to seat a stern paddler. However this would make its squared end into the prow.

The time had come for my dugout to emerge into the outside world. First, I had to move the dugout up the slope to the driveway. That happened in May, 1983. I acquired a ratchet-type log puller with wire cable and ropes. I attached one end to the canoe's stern eyebolt, and used blocks (pulleys), successively fastened to various yard trees. Slowly but surely I winched the canoe up the slope. Younger son Jason, aged 11, sat in the canoe for a bit, riding it up the slope through a sea of English ivy. Once the boat was sitting level in our parking area, I could slosh on more preservatives.

While I was doing that, the young man renting a neighbor's house strolled by with his girlfriend. He quickly examined my masterpiece and announced he was a nuclear engineer who also knew about boats. Obviously trying to impress his girlfriend, he declared my dugout unstable in the water. The girlfriend smiled with bemused sympathy. I had always assumed it would be stable, but now I wasn't 100 percent sure anymore. Of course I had not determined the centers of gravity vs the center of buoyancy.

My finished dugout needed a name. One friend later called it T-Rex. The obvious choice was turning the common noun rakìock to the proper noun *Rakìock*. Of course I wouldn't carve this name onto the boat.

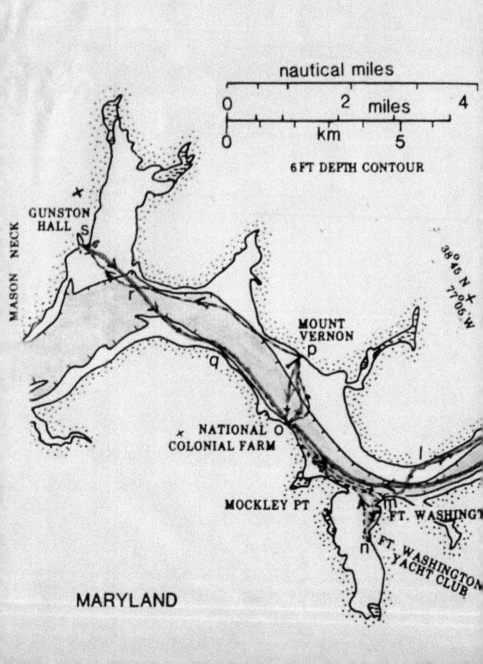

nautical miles

0   2   miles   4

0   km   5

6 FT DEPTH CONTOUR

GUNSTON
HALL
s

MASON NECK

r

q

MOUNT
VERNON
p

38 46 N
77 06 W

NATIONAL
COLONIAL FARM

o

l

MOCKLEY PT

m   FT. WASHINGTON

n   FT. WASHINGTON
     YACHT CLUB

MARYLAND

PART II

An Ancient Vessel returns to the Chesapeake Tidewater

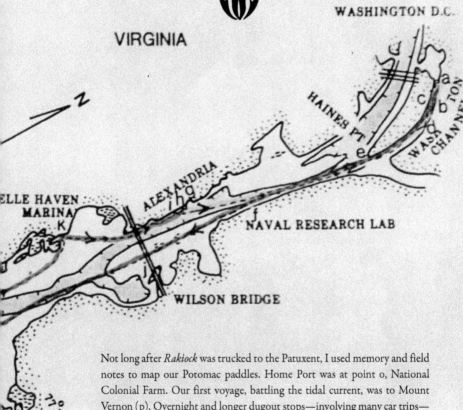

WASHINGTON D.C.

VIRGINIA

HAINES Pt

WASHINGTON CHANNEL

ELLE HAVEN MARINA

K

ALEXANDRIA

NAVAL RESEARCH LAB

WILSON BRIDGE

Not long after *Rakiock* was trucked to the Patuxent, I used memory and field notes to map our Potomac paddles. Home Port was at point o, National Colonial Farm. Our first voyage, battling the tidal current, was to Mount Vernon (p). Overnight and longer dugout stops—involving many car trips—were at points s, n, k, d and Old Town, Alexandria, where the canoe was pirated from point g by unknown joyriders, who paddled or poled it to point i. Rediscovered there with police help, *Rakiock* was secured at h. At point l, we escaped a storm and sheltered under some brush. At point f, we paddled past my workplace, where I also had recruited one co-paddler.

CHAPTER I

Scheming in Grandiose Fashion

Even before my dugout was finished in spring of 1983, I started planning for its deployment. My first choice of venue was the upper tidal part of the Patuxent River. I had previously met Rich Dolesh, head naturalist at the Patuxent River Park, on the Prince George's County side of the river. So on 23 September 1982, I wrote him one of my typically verbose letters suggesting the dugout, now "nearing the completion date" might be based at that park, used for excursions and left "on indefinite loan as one of your park exhibits."

Referring to possible historical reenactments, I wrote of having "some Colonial clothes to wear." I ended the letter saying it would be nice to have a companion along [on longer dugout trips] to cook, paddle, watch out for giant sturgeon, or just look pretty."

When this somehow did not work out, I turned to the Potomac. My thoughts now turned to National Colonial Farm across that river from Mount Vernon, a recreated farm representing late Colonial farming, created in part by the desire to preserve the historical view from Mount Vernon. I hit on a grandiose scheme to base *Rakiock* there, and with others and grant support paddle it down the Potomac from Washington, DC to Historic St. Mary's City and promote environmental restoration of the river. 1984 was also Maryland's 350th anniversary. The voyage would involve some sixteen stages, with lectures and media events at the stops. National Colonial Farm—recommended to me by a friend—would be an ideal home base for *Rakìock* also because it was only a half hour drive south from my work place, the US Naval Research Laboratory, located on the south bank of the tidal Potomac in Washington, DC.

In a letter dated 6 April 1983, I broached this idea with David Percy, National Colonial Farm director. Another idea was added: to build, under my supervision, a larger 18-foot dugout at Colonial Farm using a big rakìock that had recently been felled in the yard next to ours, struck by lightning. I suggested he might have the right contacts to get the log moved by a US Army cargo helicopter. Thinking and dreaming big was easy—especially while one is still young.

Percy's May 3 reply said the Farm "would be delighted to demonstrate [my] skills at a special weekend event as well as an ongoing activity." That was good news, although he didn't comment on my grandiose trip down the Potomac or the Army cargo helicopter. The special event was to be our first annual Potomac River Heritage Day.

In 1991, I recalled the canoe trip from our parking lot to National Colonial Farm on the Potomac. One early June day in 1983, the truck from National Colonial Farm arrived and we muscled the big hollowed-out log onto the flatbed and strapped her down. By twice weighing the truck at a weigh station on MD 301 near Upper Marlboro, with and without the canoe, we learned *Rakìock's* true weight, all of 550 pounds, or about ten aluminum canoes, and probably more than a typical native canoe of comparable dimensions.

When we arrived at Colonial Farm, all the staff were on hand in their usual 18th century outfits. (At the end of the day, the farm staff gets back into jeans and commutes home; only the caretaker family actually lived at the farm). I was kind of hoping the canoe would be put on display for the time being. What if it didn't float right and all those strangers were there to laugh? I wasn't a shipwright, after all, and I had only made salad spoons and noggins before. My thoughts returned to that arrogant young engineer–neighbor who announced that my canoe would be unstable.

But it was a warm sunny day and everyone voted to head for the river. *Rakiock* was reloaded on a flatbed hay wagon hitched to a tractor (It should have been a team of oxen, but the Farm doesn't have those—too big and brutish for our liability-conscious age). The driver in his Williamsburg-type outfit backed the wagon down a short, steep embankment onto a narrow gravel beach littered with driftwood and trash. Gently he backed the wagon into the Potomac until the water lapped over the wagon bed.

Two of us undid the straps and pushed the canoe sideways. I held my breath. Suddenly, with a 'kaplosh-plunsh' she was afloat—right side up, wonderfully stable, with a stunning 8 inches of freeboard.

The launching party splashed and swam, diving off and under the canoe, paddling a few strokes. At length it was time to return to chores and visitors. We hauled *Rakiock* up on the gravel beach and tied her to a tree.

Back then, thirty years ago, I referred to 'the *Rakiock*' as a 'she'. In later years I adopted the Navy form and dropped the article. *Rakiock* sounds to anglophones as male, so I made a linguistic sex change. As for swimming in the Potomac, we took a chance. The river water was not as clean as it is today.

In my next letter to Percy, 23 July, I 'postponed' to 1984 the 115-mile trip from DC, Fletcher's Boat House or Chain Bridge, to St. Mary's City. Admitting to getting carried away, I feverishly extended, to 230 miles, the proposed trip to Point Lookout, then up along the Western Shore of the Chesapeake Bay to Annapolis and on to St. Michael's on the Eastern Shore.

Back down to the real world, I also suggested a modest wet-run paddle from National Colonial Farm to Mount Vernon on the opposite side of the river. That would happen on a typically hot, 90ish, calm late summer day: 30 August.

CHAPTER 2

A Maiden Voyage Unheralded

Staff at Colonial Farm phoned Mount Vernon to apprise them of this imminent historic demonstration of contemporary waterborne "pickup trucks." With interpreter Wes Muller along as bow paddler, both of us in somewhat ersatz Colonial farmhand attire—including tricorn hats—we prepared to cross the historic Potomac. I tossed a home-made bailing gourd into the canoe, along with home-made paddles and a pair of modern orange life preservers. Sheep skins would cover them—to keep the modern world hidden. We untied the canoe and pushed it into the river with the stern still at the shore.

My bow canoeist climbed in first. I then pushed the dugout away from the shore while climbing in. We were underway! It was a calm, beautiful forenoon. Sitting on the carved back seat, I could both paddle and steer, occasionally correcting a tendency to yaw to the starboard because the log was a bit crooked.

We steered through glassy water green with algae, straight for the Mount Vernon pier. But once further out in the river, we noticed our actual path was upriver towards Washington. I learned an important lesson: Tidal currents have to be part of every dugout trip. I had failed to look up the tides for this day. It was obviously a flood tide. After rounding the channel buoy, we paddled very hard across and against the current, back southwest toward our goal.

Finally we reached the Mount Vernon pier, climbed out at the shore and secured the canoe. We had worked up a good appetite and hoped for some good vittles as a reward. However, there was no one at the pier to welcome us. While walking up the road to Washington's home, we encountered some tourists who asked if we were movie extras in the documentary about George Washington they heard was being filmed.

As I exaggerated in 1991, "Perhaps a little delirious with heat, we expected to be welcomed, photographed, interviewed, surrounded by beautiful women in Colonial gowns, perhaps even wined and dined. But no 'officials' appeared."

Checking in at the door of Mount Vernon, we announced to an elegant lady in Colonial attire—likely a head guide—that we were the paddlers who had come from National Colonial Farm. The first to do that in a dugout for at least a century! The lady had obviously not been apprised of our historic reenactment. So she really didn't know what to do with us. We asked if there was anything to eat. No, she said, but she would buy us each a hamburger. No thanks, we said, we have to get back. A fine welcome to historic Mount Vernon, we grumbled to each other once out of earshot.

Back down the hill and approaching our canoe, we spotted tourists seemingly peering at it. Great, we can help them understand travel on the Potomac in Washington's time. In my mind I ran through some facts to share. The tourists were looking not at our canoe. They took no note of our attire and probably assumed we were groundskeepers in uniform. We only got one question: "Is that the Potomac River?"

By then it was around high tide, and our paddle straight across back to Maryland was uneventful. *Rakiock* was a joy to paddle. Slower than a modern canoe but stable. Rides like a charm over boat wakes. Fortunately the farm office had some food in their fridge.

CHAPTER 3

Dreaming Too Big

No major canoe trips that summer. The dugout hibernated during the winter in a restored 18th century barn with hand-riven red oak siding. This barn had been donated to National Colonial Farm by Anne Arundel County. Meanwhile, ambitious plans for 1984 were evolving by correspondence, grant proposals and phone calls.

David Percy submitted (13 October, 1983) a proposal for $5,775 ($16,300 in 2022 dollars) to the Maryland Council of Humanities, with $7,445 more in 'in kind' support. The proposed October 1983-September 1984) project "A Voyage Through History: A Comparative History of the Potomac River" would be directed by Percy, with other chief scholars the historians Frederick Gutheim, Ralph Eshelman, Frederick Tilp and Ronald Brown. I was listed as "builder of historic Maryland canoes," with project duties including paddler training and voyage coordination. The 1984 trip would be in fifteen segments from southern District of Columbia to Point Lookout, and then north in the Chesapeake to the Calvert Marine Museum in Solomons. The proposal promised about 30 public events with 15 to 250 attendance at each.

In January 1984 I wrote a separate proposal to Serge Korff, the physicist who was a mainstay and former president of The Explorers Club, of which I was then a Fellow. I requested $1,450—over $4,000 in today's money—to defray the cost of photography, auto travel to and from canoe stops, and subsequent voyage narrative publication.

For help regarding tidal currents en route down the Potomac and beyond in the period 1 June to 1 September, Percy mailed Paul Wolff, assistant administrator of the National Oceanic and Atmospheric Administration (NOAA).

Neither of the grant proposals succeeded. The call for volunteers had little response. The Explorers Club sent a form rejection notice without comments. I eventually dropped my membership. Too many wannabe, pseudo and arm-chair explorers. Scientists are used to proposal rejections, but not via boiler-plate forms.

By the spring of 1984 the idea of a grand dugout voyage down the Potomac was downsized to a paddle from National Colonial Farm to the annual Alexandria Red Cross Waterfront Festival, 2–3 June, 1984, and the following weekend on to the Potomac Riverfest '84, at Hains Point and the southwest DC waterfront.

Percy contacted the organizers of the Alexandria festival. "The participation of Peter Vogt and the National Colonial Farm log canoe" was formally requested in a May 1 letter by B.C. May of the American Red Cross, the festival chairman. May's letter advertised "tall ships, small ships, exhibits, music, power boating, board sailing, and entertainment on two stages." We were advised to "arrive at the North Pier in Alexandria around 12 noon in time for the opening ceremony." The flier for Potomac Riverfest '84, signed by DC Mayor Marion Barry, celebrated the environmental "rebirth of the Potomac River and honoring those who helped make it possible."

CHAPTER 4

Wind and Waves

Now we needed a longer wet run to prepare for padding *Rakiock* to festivals. On May 5, 1984, at 10 a.m., interpreter Greg Starbuck and I set out upriver towards Fort Washington Park and from there into Piscataway Bay to the Fort Washington Yacht Club Marina, where some sort of opening ceremony was planned later that day. We took on: 3 paddles, rope, 2 sheepskins, period dress (+ sunglasses, watch), covered imitation wooden bucket container w. food, knife, water. Greg brought a bundle of fliers to hand out—about National Colonial Farm and about dugouts. He also brought his Colonial-type pipe, which he had learned to play.

Here is the journey as recorded in my unedited trip log of the festival voyage (five stages total).

The wind was initially from SW, about 15 knots, waves 1 ft, few white caps. Somewhat rough, marginal

Our canoe was trying hard to turn into the trough, which is also a problem for regular canoes and ships in storms took on a little water on 4 or 5 occasions when canoe inadvertently turned [counter clockwise] . . . into trough . . . canoed under pier [fishing pier next to NCF] . . . within 50-100 yds of shore . . . At Mockley Point ash trees surrounded by water, canoed inside one tree!

Mockley Point is a rich archaeological site, giving its name to Mockley Ware, a Piscataway pottery-making style. By then it was about 11 a.m. with the winds and waves subsiding. We paddled across the mouth of Piscataway Creek to about 200 yards east of the Fort Washington light.

A large freighter passed but [we] took on no water from wake . . . Paddled against 0.5-1 kt current around point (& fishermens' lines) . . . 12 noon, Pulled stern around up and slightly on sand . . . Picknicked, Passed out a few fliers . . . Walked up to chat with park ranger, bearded bulky character . . . whose initial remark was to get beard & mustache off because not worn in colonial times

After touring the fort, we walked down to *Rakiock* with the park ranger. He suggested a Civil War use of dugouts in this area. Union troops deserting in dugouts under cover of night were apprehended and thrown in the Fort Washington brig. Driving back to the fort the next day, we heard the ranger's superior suggesting fugitive slaves fleeing by dugout would seek asylum at the fort. A distant dugout certainly resembles a driftwood log, provided the paddlers crouch down. Maybe that helps explain why Chesapeake natives did not decorate their dugouts.

Meanwhile, we pushed off about 2 pm . . . Tide had reversed, so we were paddling against the current again! But currents slack off rapidly inside Piscataway Creek . . . Greg took a few pictures enroute . . . will add some flair on future trips . . . Canoed along heavily wooded steep bluffs between Ft. Wash. & marina . . . Dogwoods in bloom! . . . Beached canoe for another 45 min of looking at broken concrete ?? lookout station? . . . & walked along pebble beach . . . Found a few Indian scrapers (?) . . . Good exposure of Cretaceous (Patapsco??) multicolored clays on bluff just west of marina . . . Reached marina at 4 pm . . . sat through opening ceremony at yacht club . . . Greg Starbuck got up to make a short announcement . . . Had some punch & cookies at reception . . . Several people asked about canoe and took pictures . . . Greg handed out fliers . . . Drove back to Farm in Greg's car about 6:30 . . . Drove directly to Randi's office party at Sally McGrath's

The return paddle to Colonial Farm would take place a few days later.

Canoe left tied up at Marina until following Thursday (May 10, 2- 4:30 pm) . . . Attached one line to piling low, so that when we canoed back to farm at high tide, it was too deep in the water, and I had to cut it off . . . Dr. Percy drove Mary Catherine Butler & myself to Yacht Club to take canoe back . . . Had stopped there earlier to bail out 2" rainwater . . . Julie Coyle who manages the Yacht Club took a photograph of us leaving . . . should contact for a copy, only picture of this first "co-ed" trip . . . Two hour return smooth, uneventful . . . No water into canoe . . . Used ratchet log puller to pull canoe up on bank

There was one more National Colonial Farm dugout event, on May 27, before the festival voyage. However *Rakìock* would not be launched but only shown to visitors. It was a show-and-tell by Greg Starbuck and me. I had previously arranged for a large donated rakìock log from a farm west of the Patuxent to be hauled to the farm, a blank on which dugout making would be demonstrated. To groups of twenty to thirty, we demonstrated sawing vertically to the gunwale line and then knocking off blocks or slabs. At one end of the log, the grain required sawing off a slab horizontally. We used dogwood gluts I had made from a windfall tree near our house.

We would be National Colonial Farm ambassadors and took along many fliers to hand out, plus some special Farm herb bouquets as gifts.

1 June 1984 (Friday) After bad weather forced delaying one day after the other starting Tuesday, Friday was the last

chance to make Belle Haven Marina in order to reach Alexandria's waterfront Festival the next day at noon. (Saturday). Actually, under calm conditions and incoming tide, the trip from Colonial Farm to Alexandria could easily have been made in one day.

My log next summarized the trip to Belle Haven before returning to trip details.

As it was, we (Norman Cherkis of NRL and myself) left the Farm at 1:15 pm Friday, and arrived at Belle Haven Marina 6:45 . . . This was with unfavorable tide and at times very marginal wind & waves! . . . This amounted to somewhat less than 6 hours on the water, excluding two close-spaced stops on the Virginia shore about ½ mile north, opposite Ft. Washington . . .

We had passed Mockley Pt 2 pm . . . and Ft. Washington 3:15 pm . . . Slightly north of Ft. Washington we turned NW to cross the river, which took about an hour . . . After a 20 min rest stop we continued about 15 min along the shore but once more beached our canoe (4:30–5 pm) and sat out a thunderstorm (the edge of one, not more than 0.25" rain and mainly cloud-to-cloud lightning) . . . The thick vegetation along river bank, overhanging the narrow pebbly beach provides pretty good shelter against rain!

My log now returns to recount the wind and waves encountered starting at National Colonial Farm, and how the dugout performed.

When we left the Farm the wind, nearly calm at dawn, had picked up to 15–20 kts and there was a 1–2 ft chop on the river with scattered whitecaps . . . I was apprehensive, to say the least!

Fortunately the wind and waves had a component in our direction of travel, so we made reasonably good time. However we took water overside at intervals right from the start, splashing over the bow or lapping over the gunwales. Within minutes of the start, everything was wet! Just past the pier, Norman threw one bailing gourd back and it flew overboard, but I managed to spear it with my paddle. We stopped several times to bail out about 1" water (average) although this much water did not seem to have any significant effect on handling or stability.

There was a terrific force turning canoe broadside (CCW) and both paddling on left side could not prevent turning. Often I did little more than steer, levering the paddle against the side of canoe and sculling hard. It may be that this particular canoe, by not being exactly symmetrical about a vertical lengthwise axis, also has a natural tendency to turn left in the absence of wind and waves.

By the time we reached Ft. Wash the wind was down to 10–15 kts and waves to 1 ft. However the wind had turned from WSW to NW. It was initially sunny but there was increasing cumulus buildup leading to the thunderstorm mentioned. On the transit across the Potomac the thinner of the two poplar paddles snapped (paddle made by Les Muller or Doug Lockhart, NCF staff or volunteers). Tulip is obviously inferior for paddles!

Before the storm we encountered stiff headwinds (N at 15 kts or so) & it got rather chilly. The sun came out at about 5:15 and the wind died down completely, The last hour was like glass, the afternoon sun illuminating cumulus clouds. After passing near shore estates, some architecturally inappropriate, we canoed straight across the broad bay north of Bellehaven . . . Saw a few fish jumping and a heron three times on trip (or three herons once each).

In my memory nearly four decades later, we raced triumphantly at 2 knots through a flotilla of small sailboats, all becalmed. However this is not mentioned in my log, so it might have been later wishful thinking.

Belle Haven young man responsible for boats was helpful & friendly . . . several sailboaters on hand expressed interest in canoe . . . Tied up overnight along floating dock . . .

Norman called his son . . . eventually got through . . . ride to Cherkis house . . . then back to Farm to pick up my car.

All in all, this so far longest trip (& most strenuous) exposed us & dugout to a wide variety of wind and waves, plus an adverse tide the whole way (However since we were only in the main channel 1 hr out of 6, the average adverse tidal current may only have been a few tenths of a knot.) The canoe rides the waves rather well, and probably is superior to a regular canoe in gusty wind. There was never a danger of rolling so far over as to take say a bucket of water or more at a time, and certainly no danger of capsizing (if in fact this kind of boat can be capsized).

However for waves above 1 ft and winds much above 10 ks with a 1 mile fetch, the inside of canoe will get wet and proper packaging of materials is essential . . . Norman carried dry clothes in plastic sack . . . but they got wet on account of small hole in sack! . . . Camera also got wet inside Ziploc bag, but this must have been condensation.

Since water was warm (high 60s?), & weather in 70s, it wasn't that bad to be wet, but a trip in the cold seasons would be unpleasant!

To us inexperienced canoeists not in perfect form, the main complaints were some soreness in shoulders, biceps and lower back and mid-section muscles. I bruised the outside of my thumb by pinching it between paddle & sides as well as inside of arm above wrist due to rubbing against unfinished gunwales, or outsides of canoe.

Will have to experiment with loading of canoe with cargo, e.g., a collection of farm cargo across to Mount Vernon. Obviously more green water will come over the sides if the boat is heavily loaded. Furthermore a single canoeist could handle a laden 15-foot dugout, but only under calm water, favorable or no tide/current conditions. [The nautical term 'green water' does not refer to color, but to distinguish it from splashed water, called 'white'.]

The trip to the Alexandria festival had to be continued on the forenoon of the next day.

Saturday 2 June . . . Arrived at Belle Haven Marina at 9 am . . . weather clear, warm (70s), but breezy . . . (10–30 kts from NW) . . . Departed 9:10 am for trip across broad bay toward Woodrow Wilson Bridge . . . Passed small, submerged islet with a few ash trees . . . Gusty wind, waves 1–2 ft again from the port bow (NW) . . . again took on some green water & had to bail occasionally . . . Very confused sea where waves from Cameron Run Bay interfere with waves coming straight down from north . . . Passed under Woodrow Wilson Bridge at 10:10 (one hr) & after some more tough paddling against chop and wind, reached yacht club at about 11 am . . . The manager, a blunt unpleasant type, told us to beat it! No chance to keep dugout here . . . perhaps I had misunderstood the Festival manager BC May . . .

Canoed around USS *Hoist* (SV) & *Lindoe* and came into dock next to *Hoist* . . . Greeted by Coast Guard Reserve officer in long beard & a good looking 50ish lady in white Commodore uniform . . . Ms Riggs showed us her houseboat . . . She had worked at Scripps & knew oceanographers at Lamont. Norman and I traded sea stories with her . . . Had coffee & donuts . . . canoed back to Lindoe christening, where a very small crowd listened to festival potentates congratulate themselves & an old lady tried about 8 times to break champagne bottle . . . [*Lindoe* was a Swedish tall ship (schooner) purchased by Old Town and renamed the Alexandria.] . . .

. . . BC May, the chief organizer, didn't work us into the ceremony, and he seemed perplexed about how we should be demonstrating the canoe . . . Handed out a few fliers, took in festival, beer & food at Founders' Park . . . Still very windy . . . motorboat races canceled . . .

wind surfer exhibition canceled . . . How did we ever
get here in our clumsy craft? . . . About 2:30–3 pm . . .
moved canoe back around and hid it under floating
ramp of Dept of Education pier . . . Walking back we
stumbled on TV crew, gave away herb bouquet &
blurted something into microphones . . . Hope they
didn't air it! . . . Was caught unprepared . . . Cherkises
drove me back to Belle Haven . . . took car back to
festival . . . wonderful chat with Tilp & woodcarver
(Swartz) & listened to our own [Tom] Wisner—who
wasn't sure he knew me . . . Fireworks at 10 pm &
headed home . . . Will need to sneak into boatyard by
climbing around fence to get canoe next Saturday.

CHAPTER 5

Rakiock Goes Missing

I updated my log on June 10 after returning home from the second day of Potomac Riverfest '84.

> Sunday other responsibilities . . . did not visit Alexandria . . . but some friends said they saw it . . . Others saw a bit on 6 pm Sat Channel 4— fortunately not the interview.

> Mon. 3 Jun 6:30 . . . drove to Old Town after work . . . Climbed around fence (Founders Park side) for a look . . . Horrors it was gone! . . . Despair . . . all the hours invested . . . all the wonderful plans!

As I recalled in 1991 "If it had been cut adrift by vandals it could be anywhere by now, just a floating log mixed among driftwood."

> In a bleak mood . . . Walked over to the Old Dominion Boat Club . . . one of the ? officers of club name ? was very sympathetic . . . said he'd announce to members . . . He gave me [phone numbers of] Coast Guard, Harbor Police and Alexandria City Police.

"You lost a what?" [I recall also being asked for its registration number] Harbor Police said they have no jurisdiction on river inside pier lines.

> Alexandria cop came . . . I showed him site . . . he made report . . . then we walked along waterfront (north) & found the canoe beached inside old rusty pilings . . . Robinson's Terminal north of Founder's Park . . . (Phew!).

My 1991 account was more comprehensive:

The Alexandria cop arrived and with a skeptical smirk, began to fill out a report on the theft. I'm sure he didn't realize that canoes were considered so important in Colonial times that a law was passed which made the unauthorized taking of a canoe a FELONY. He and I walked along the waterfront, looking for the canoe. Finally I spied a dark log off in the distance, among rusting pilings of the Robinson Terminal north of Founder's Park. Unfortunately we couldn't get to it, since the gate was padlocked.

My 1984 field notes continue:

It was inaccessible . . . so we walked to the 1 Duke Terminal through warehouse full of giant paper rolls being unloaded from a Swedish freighter . . . found a forklift operator who hailed the manager who was on the ship . . . The three of us (cop, manager, me) drove to Terminal, unlocked gate . . . inspected canoe . . . Full of mud tracks, a board, a stick, lines in knots, tied to a stone ashore . . . Some youngsters joy ride, the little *(&!

Cop dropped me off at Boat Club . . . Manager said he'd leave gate unlocked . . . Nobody at Club wanted to help me get it . . . It was getting dark . . . meanwhile. Asked some joggers, bicyclist, then approached 3 Black teenagers playing handball over a volleyball net . . . One came along . . . hadn't ever canoed before but eventually caught on . . . Got it to Old Dominion . . . ironically the last all White male boat club in the area

(In my 1991 blurb, I noted that the joy-riders had beached the canoe at high tide, so much of it was resting 'high and dry' on foul-smelling mud. It took both of us to push the canoe into the river, and we sank into the mud in the process.)

First tied up next to dock (non-floating) . . . Thought better of it . . . canoe drifted under dock & would have got mangled at high tide . . . Tied up thus [log includes sketches] & used slip closest to land . . . next to little park, which had been expropriated from the Old Dominion Boat Club . . .

To tie the boat that way I had to jump into water & wade, throwing shoes over first to surprise a park bench couple . . . Left Farm fliers . . . reparked . . . had 2 beers at Fish Market & home . . . An unplanned evening!

I drove from work to Old Town again on June 7 . . . Checked on canoe . . . Thurs. was ok . . . Met City archaeologist . . . Ms. ?? was interested doing some filming canoe to help show how river was . . . Should follow up!

CHAPTER 6

The Potomac River Festival

My log continues:

> Met Cat Butler [Mary Catherine Butler of National Colonial Farm] at Jefferson School near Gangplank Rest . . . It was already hot at 9 am! . . . We left her car there & drove to Old Town . . . Shoved off at 10:10 . . . River rather calm, 5 kts. SW to W wind . . . Passed just off NRL pier & drifted for lunch. Tide slowly upriver . . . Stayed close to Bolling shore . . . More & more wakes, boats, incl. large barge . . . No water over gunwales though . . . Despite north flowing tide, encountered southward currents around noon south of Anacostia mouth & made slow progress . . . Motorboat came over and gave each of us a beer . . . Crossed mouth of Washington Channel at Hains Point & canoed along Hains Pt. side . . . Parade of Boats was just exiting Channel . . . Many wakes & reflected wakes, & finally took in some water.
>
> We got our pictures taken by boaters · · · It was HOT!
>
> Cat is uncomplaining and good to have along · · · We chatted with Harbor Patrol in mid-channel (one was from Louisiana & said pirogues still used there! · · · Took on some iced tea & then tied up at metal dock, erstwhile home of the *Sequoia* · · · Climbed up in bare feet & OUCH! · · · Collapsed in shade sipping Cokes · · · It was about 2:30 · · · Again we had averaged about a knot.

[*Sequoia* was the presidential yacht, built in 1925, and serving nine presidents from Herbert Hoover to Jimmy Carter. In 1977 Carter ordered it sold to emphasize reduction in US government cost.]

Almost no-one there at the appointed 'launching' 4 pm · · · First, the brochure hadn't said where · · · second, ferry from other side had stopped running at 2 pm · · · Instead of 200,000 or 40,000, there were maybe a few thousand at fest · · · perhaps the heat · · · Disorganization rampant · · · nothing on schedule · · · We paddled to other side · · · had some free beer · · · tied up at Commodore Ms Marcia Crossley's dock (Capitol Boat Club) & left it there overnight · · · Paraded up and down walkway a few times & back to Alexandria to get my car & then home.

SUNDAY JUNE 10

Next day, hot again (temp in mid to high 90s both days) · · · met Augie [August Selckmann, Calvert County 'hippie', canoeist, and professional photographer] at school at 10:30 · · · Took Mutsie (Augie's dog) along of course · · · Canoed over to gate at entrance to Tidal Basin, back to metal dock · · · Tied up, walked all the way to Hains Pt. itself · · · back in time for 'demo' · · · A few people here & there, but good interest, contact · · · Handed out a few fliers · · · Took fat, aggressive California "girl" for short ride, nearly capsized us · · · Then Augie & I canoed close to shore at request of somewhat plastered [fellow of 20 or so] · · · He asked to take ride & jumped in before I could get away · · · Then he dove in from dock, bringing police over in a hurry · · · He climbed ashore across the canoe soaking our sheepskins, complaining about his wet cigarettes.

We paddled over to Gangplank · · · had a delightful
2 hours drinking courtesy beer with the management.
· · · Augie took some of their teenage girls out for
short spins · · · Dusk came · · · We canoed in & out
among slips, chatting with boaters · · · "We're spirits
of Potomacs past" & taking on drinks. · · · Gave two
rides · · · Handed out fliers by sticking them to wet
paddles · · · An outraged squeal told me I had picked up
Mutsie [instead of the sheepskin the dog resembled and
was lying on.] · · · Watched super fireworks against full
moon from dugout · · · Tied up at Gangplank Marina
for the week · · · Oh yes, about 4:30 got videotaped by
crew documenting festival for mayor (Marion Barry).

The dugout was paddled back to NCF nonstop · · · 14 miles · · · on Wednesday, 20 June the average speed was around 2 ½ mph · · · probably with some help from ebbing tide · · ·

Farm 9 am · · · John Brozena (scientist colleague at Naval Research Laboratory) · · · Susan Rhoads (grad student working on her MS under my supervision at NRL) 9:30 drove us to Gangplank.

11:00 lv Gangplank Marina

11:40 Hains Pt.

1:10 NRL Pier

(In 1991 I wrote that our passing NRL was greeted by 'the waves and shouts of office colleagues'.)

2:10 Woodrow Wilson Bridge

4:30 Fort Washington

5:45 Colonial Farm

5:45–6:45 ratchet up & secure canoe

Weather warm, relatively low humidity, high 80s · · · scattered variable altocumulus & cirrus · · · winds 0 to 10 kts north backing to west · · · Few boats on river · · · DC Harbor Patrol cruised up to check us for life preservers · · · Hains Pt. · · · Wash Post or Times nature writer took flier · · · said he would get in touch.

Later that summer and back at National Colonial Farm, I participated in several farm events and did a bit of experimental archaeology.

ONE FARM EVENT:

11 August Corn Harvest Festival · · · canoe in water 12 noon–4 pm · · · Gave rides to 1–2 kids [per trip] with & without accompanying adult · · · Picked them up at new pier, went ca. 100 yards & return · · · Fun! · · · total ½ mile.

I was curious about the cargo capacity, the total weight besides myself. This, of course, depends on the density of the cargo and how it's stowed in the boat. Experimenting on very short trips, I loaded the dugout with the type of cargo that might have been carried in the 18th century. On the first such trial, I carefully stacked 340 pounds of Colonial-type bricks (five dozen). Adding myself, paddles, sheepskins and four small pumpkins this came to 655 pounds. After paddling my cargo barge around near Colonial Farm, I returned to shore and added about 315 pounds of sacked corn. At that point, about 700 pounds of cargo, not counting my nearly 200 pounds, the dugout became a bit wobbly. Still more, and *Rakiock* might have rolled to the side and spilled its cargo—and maybe even me.

There was no problem paddling alone or with a non-paddling passenger facing and talking to me. I almost always served as the stern paddler and sat on the comfortable seat. When *Rakiock* was heavily loaded, I sat on the bottom to improve stability

We always had a deer pelt covering the PFDs, and some smaller sheepskins and furs for comfort and show. On some trips, a bale of straw worked as a backrest for the bow paddler. We paddled with home-made paddles, one of hickory and one of sycamore. A modern, lighter weight, laminated wood paddle was often carried for emergencies. A long homemade hickory pole with a small fork at the tip was carried for poling in shallow water. We carried water in an authentic ceramic jug, and a wooden container held lunch.

The likelihood of future paddling, also on the Patuxent and Chesapeake, meant repeatedly loading *Rakiock* onto and off of trucks, and moving it about on the ground. I wrote up and distributed my written and sketch-illustrated guidelines on how to do that without gouging the dugout. These guidelines are undated and may have been written already in 1983.

IMPORTANT–PLEASE READ

Tips on Handing & Moving Dugout so I can sleep soundly—have invested several 100 hrs!!

Tulip poplar wood is relatively weak & soft & the dugout is heavy. If much of its weight is supported on sharp edge or point, it will gouge holes in the wood. These will catch on things later & will rot first.

Do not pull boat across, or support it on, or drop it on, a sharp wood or metal edge

Also please do not put a chain, cable or grappling (logging) hook around or into it at any time.

I have found one person can move it slowly with a flat strong board (6 ft 2x6, say) & a fulcrum log, by pushing down (levering) & turning at same time. This method is good for turning dugout around on ground.

Two people can easily pull it on level ground using rollers (best 4 ft long 4" diameter and nice cylinders) with circular x-sections and no bends., May be good to make about 5 for use now and next year.

Helps to oil rollers so if they get caught, the thing will slide. I used motor oil, but grease might be better.

To raise the dugout to a higher level, lever it up with a flat board (first one end, then the other) and place a sack of flat boards or other flat stable things underneath, one after another.

For levering it up, two people should work (from both sides) so dugout will not slide sideways. A third person moves the supports. Watch out for fingers. Do not use crowbars, metal pipes, etc. as levers & these techniques (others we need to work out) will be especially important when we get a whole log that will weigh in at 1 to 1 ½ tons.

If necessary, probably 8, better 10, people, in pairs, using 2x6s, could carry a canoe.

It could also be carried upside down by gripping gunwales. But do not pry it or lever it in upside down positions, it will break or gouge the gunwales.

If come-alongs or log-hoists, etc., are used, remember that this gives a person enough mechanical advantage to damage the canoe. Do not pull across sharp edges or across any large stones or metal structures.

Don't pull it around a corner. Attach come-along only on eye bolt. Do not move by wrapping a cable or rope around dugout.

When moving on hay wagon, might be useful to take two 10' logs or beams and make cradles to protect from sliding off wagon when on incline.

Our furthest southward foray was Gunston Cove, Virginia. Before setting off there, I first drove to historic Gunston Hall and obtained permission to land and keep the dugout for some days in a small inlet near the mouth of the Cove on its eastern side. This inlet, also called a 'canal', served around 1900 as the place where J.P. Morgan had anchored his yacht. It's just a short hike from George Mason's estate.

HERE WAS THE GUNSTON COVE EXPEDITION:

THURSDAY, OCTOBER 18:

12:30–1:25 pm · · · 50–55 minutes National Colonial Farm–Mt Vernon Pier · · · canoed alone · · · Slight incoming to high tide · · · 3"–6" waves from north · · · light winds, partly cloudy to sunny, hazy, warm (70s), water low 60s · · · First trip alone

(I stopped there to pick up the two other canoeists, so this would be the only longer trip with three paddlers.)

2:15-4:20 pm · · · Mt Vernon to Gunston Hall 'canal' mouth · · · Canoed with 3 canoeists (Greg Vink, John Trieul and myself · · · Vink was my post-doctoral fellow associate) · · · Calm, glassy water to slightly rippled · · · weak to moderate south-going tidal current · · ·

Hydrilla patches, driftwood, some with gulls on it · · ·
Only 3 boats passed · · · Somewhat tippier with 3, but
no mishaps · · · No water taken.

4:30–5:30 pm · · · explored 'canal', beautiful small
wooded inlet, autumn leaves · · · Got canoe stuck on
submerged log, twice · · · had to get into water to
stand on log and pull it off · · · Greg broke his paddle
prying · · · Tied up canoe just inside cove next to bank
· · · 1 mile walk to Gunston Hall through fine old oak
forest · · · They were preparing for "Regents" dinner &
were not happy to see us.

SUNDAY OCTOBER 20

Jason & Randi & I drove to Gunston · · · canoed a little
bit in the 'canal' & once briefly out into the Potomac
· · · Warm fall weather · · · total ca ¼ mile · · · Some
photography of inlet and dugout.

SUNDAY OCTOBER 27

Another warm day · · · Called 'Buf' Huntley (National
Colonial Farm) from NRL · · · canoed 2–4:30 pm · · ·
out on river +/- ¼ mile from inlet · · · Grounded on
way in low tide · · · had to drag canoe · · · excavating in
front of it with paddle · · · Ca. 1 mile total.

Rakiock was left secured in that Gunston Cove inlet for
another month. It was getting late in the year, with colder
water and more problematic weather. However I had to get the
dugout back to NCF. In late November I paddled the dugout
back alone:

30 Nov · · · tide rising · · · 10:30 Gunston · · · clear,
bright sunshine, few cirrus · · · Fog dissipating · · ·
winds light SW 5 kts to calm · · · waves 0–3" · · · temp
35 deg. start · · · 50–55 deg. end · · · Water cold, clear
in places 5 ft bottom clearly visible · · · Few water birds
· · · 1 or 2 boats · · · Very enjoyable! · · · After dragging
canoe out of Gunston 'canal' (used waders) Preston
drove me back to Gunston to get car · · · (36 miles NCF
to Gunston (1 hr) vs 5 miles by water (2 hrs).

PART III

Splashing the Patuxent

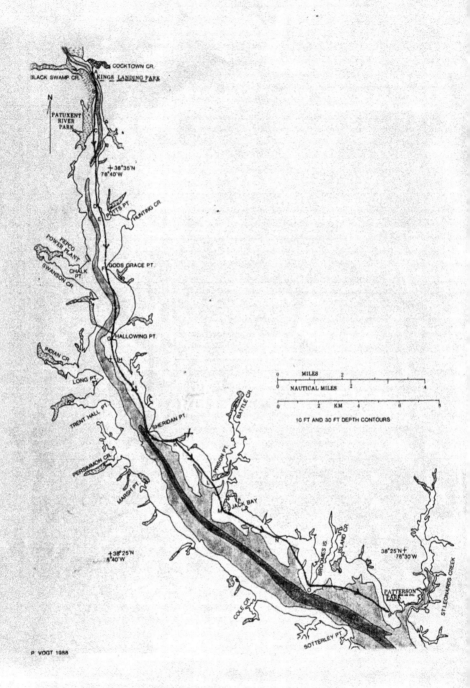

COCKTOWN CR.

BLACK SWAMP CR.

KINGS LANDING PARK

N

PATUXENT
RIVER
PARK

+ 38°35'N
76°40'W

PITTS PT.

HUNTING CR.

PEPCO
POWER PLANT

CHALK
PT.

SWANSON CR.

GODS GRACE PT.

HALLOWING PT.

INDIAN CR.

LONG PT.

TRENT HALL PT.

SHERIDAN PT.

BATTLE CR.

PRISON PT.

PERSIMMON CR.

MARSH PT.

JACK BAY

BROOMES IS.

ISLAND CR.

38°25'N+
76°30'W

MILES   2

NAUTICAL MILES   2   4

2   KM   4   6

10 FT AND 30 FT DEPTH CONTOURS

+ 38°25'N
76°40'W

COLE CR.

PATTERSON
PARK

ST.LEONARDS CREEK

SOTTERLEY PT.

P. VOGT 1988

CHAPTER I

Point by Point, a Journey

In 1985, the dugout was trucked from National Colonial Farm to King's Landing Park in Calvert County in time for the first Patuxent River Discovery Day. Field notes, if any, did not survive, but a May, 1985 *Calvert Recorder* article about the event included a photo of me sitting in dugout just offshore, evidently waiting for passengers

Its next recorded deployment was an appropriate one: a regional Boy Scout Camporee featuring Native Americans.

> Saturday, October 12, 1985 · · · Cool bright calm to light winds · · · Short trips with various Scouts at So Md Camporee · · · probably 25 different Scouts got a ride · · · total 1 mile. · · · Stored at King's Landing first in barn then in concrete restroom/stable blockhouse south of other bldgs.

I recall one of the paddles in which my two passenger Scouts, upon hearing me assert that this dugout could not be capsized, tried to prove me wrong. They both jumped on the gunwales. *Rakiock* heeled over hard but did not roll over.

The next recorded event including the dugout was the following year—Patuxent Discovery Day.

> Saturday, May 3, 1986. According to the flier, this was an Open House, 9 am to 4 pm Numerous events and exhibits, starting with 10 am DNR Hovercraft on the waterfront. Tom Wisner and Mary Sue Kaelin were featured (Music and Storytelling) from 2–4 pm. MD Governor Harry Hughes completed the events with his 3–4 pm address "The Patuxent River." Rakìock rides were featured under "Boat Trips on the Patuxent River-Every 30–45 minutes" as "Log Dug-out Canoe-Canoe Indian Style."

I used the back of the flier to record my field notes:

> Cold, sunny, very windy, 50s. NW winds, very low tide. Launched dugout for two short rides only...total ⅛ mile

A 21-mile dugout voyage—from King's Landing Park to Jefferson Patterson Park and Museum (JPPM) took place on 16 and 17 October, 1987. It was experimental archaeology 101. Wayne Clark, archaeologist and JPPM director, was bow canoeist and kept a careful enroute log, the best record of dugout *Rakìock*'s performance and of what we saw and did. His log was finalized in 1988: Official Log: Patuxent River Log Canoe Experimental Trip Number One.

We broke the trip into two 10-mile segments, with an overnight stop parking the canoe on the river bank below the historic farm of Judge Perry Bowen and his wife Virginia. We planned to depart each day around high tide, so the ebb tide would speed us up near the start. Fourteen waypoints identifiable on maps were chosen (A for King's Landing Park and P for Jefferson Patterson Park and Museum), and Wayne calculated average speeds for each segment and for the entire trip.

Following are excerpts from his 1988-edited log. The first person refers to Wayne Clark, who paddled from the bow, while I paddled and steered sitting on my stern seat.

Logistics: I arrived at work at 9 am and drove the truck to Peterson Point where I left it for the purpose of driving us home the next day...Stu Reeve took me to King's Landing... the two maintenance men at King's Landing had already managed to have transported the canoe from the barn to the water's edge, an almost superhuman task for two people given the 600-lb weight...

Point A: King's Landing. We left King's Landing at 10:25 am. The first few moments after launch were spent getting used to the balance of the canoe. It tended to tip but would not overturn...I sat in a cross-legged or extended leg position in the bow...with a bag for a back rest.

Point B: Power Line Crossing. We worked our way to the Prince George's side of the river to look at the extensive freshwater marsh. We saw one blue heron on a duck blind...A number of duck blinds lined the shore...the winds were from the north at 5 knots...The Chalk Point Power Plant came into view and diminished the view for a good portion of the trip today...detracted from the natural beauty of the river. A to B distance 0.6 miles.

Point C: Marsh Opposite Deep Landing. Arrived 11:15 am after following the western edge of the river along the marsh. A power boat heading down river passed us, the first boat of the day. This part of the river is part of Patuxent River State Park. We stopped here for a break from 11:15 to 11:20 am and decided to head to the Calvert County side of the river...B to C distance 1.23 miles.

Point D: Point North of Potts Point. Arrived at 12:20 pm. This was a long stretch of the river but the tide had begun to run out. When we stopped paddling, the canoe drifted downriver at about half a knot. We wanted to stop at this point after the long stretch but found this area offensive because of an electric generator being used to produce power for construction of a new house next to the point. It was amazing how distracting this noise was compared to the silence of the river. We noted a number of new houses going up along Potts Road and the River front with much more development on the Calvert County side as compared to the Prince Georges County side... After we passed this point we decided to have lunch, so we stopped paddling and ate lunch while the outgoing tide took us downriver. C to D distance 2.3 miles.

Point E: Potts Point. We arrived at Potts Point at 1:00 pm. D to E distance 0.5 miles.

Point F: God's Grace Point. After crossing the mouth of Hunting Creek we arrived at God's Grace Point at 1:40 pm. This was a very attractive sand beach with marsh behind the beach, and somewhat isolated from development. We decided to beach the canoe to stretch our legs and look around at the beach. We noted one fisherman with two fishing poles at the tip of God's Grace Point. The fisherman turned out to be Judge Middleton, Orphans Court Judge for Charles County. He owns a farm in Charles County called Cedar Hill Farm. He stated that when he cleared the trees in the 1930s around one spring he found a rich Indian site which has produced over a barrel of projectile points, many of which he still has. This site is still under cultivation and produces an occasional point. I joked about him fishing on the Calvert County side of the river and he said that many of his favorite spots on the west side of the river were closed after purchase by DNR for the Patuxent River Park. He had been at God's Grace Point all morning and had just caught his first fish of the day at the tip of the Point. We departed God's Grace Point at 1:50 pm. E to F distance 1.5 miles...As we traveled down the river, we were passed by a clam boat going upriver and noted that he came back down river at 2:10 pm. The boat we saw going downriver earlier in the day came back upriver. The canoe rode the waves from the two boats with ease, receiving no water in the canoe from the waves.

Point G: Benedict Bridge, Route 231. We arrived at the bridge at 2:35 pm after traveling in an approximate straight line from Gods Grace Point to Hallowing Point. We noted four fishing boats under the bridge and when we questioned one boat about the day's results they stated that fishing was "slow." The wind during this stretch was approximately 5 knots from the south, resulting in a rippled surface but not providing much wind resistance. The wind started up at midpoint between F and G so we decided to increase our effort for fear that the stronger winds predicted by the weatherman would soon be upon us. We were pleased to be beyond the sight of the power plant, power lines and bridge, and looked forward to seeing the more natural view of the lower part of the river. We did not stop at the bridge [except to pee] but proceeded paddling to reach the unnamed point north of Buzzard Island. F to G distance 2.0 miles.

Point H: Point North of Buzzard Island. This unnamed point is an undeveloped area and very nice. We arrived at 2:55 pm. The wind decreased to about 3 knots with very low waves and the tidal current was still strong. As we passed

Hallowing Point we noted the extensive shell midden site which had been disturbed by modern homes and bulk heading. I think this is the site tested on various occasions by Steve Israel and first found by Richard Stearns. We did not stop at this point but continued onto the next stop past Buzzard Island. G to H distance 0.9 miles.

Point I: Open Water Opposite Sandy Point. Arrived at 3:25 pm and took a five minute break (3:30). From there we moved toward shore to look at eroding bank which was 20 feet tall but we did not get real close to shore. H to I distance 0.9 miles.

Point J: Sheridan Point. Arrived at 4:04 pm. We saw two small boats on this leg of the trip. The winds were calm and the tides still going out perhaps at a slower rate. The Sheridan Point mansion is quite striking from the water. We were growing tired by this point. We stopped for a break to walk around the point. The clam boat passed us again and so did a pontoon boat. I noted that the very point had quartz, quartzite and some jasper cobbles about one half to three inches in diameter in only a 20 ft area...rest of beach had small pebbles. I noted one fire-cracked rock and Peter found one quartzite core which was retained. The core was found about 400 feet east of the point on the north shore. We departed at 4:45 pm. I to J distance 1.5 miles.

Point K: Bowen's Beach. We arrived at Judge Perry Bowen Beach at 5:20 pm having run aground in the area of the eroding clay deposits between the beach and Sheridan Point. We noted a series of small sites eroding out of the bank along this section including an extensive shell midden in the Bowen house yard. As we knew that we were running ahead of schedule we took our time along this leg of the trip. The water was perfectly calm

with no wind. We were at slack tide. It took until 6:00 pm
to pull, push and wedge the log canoe up onto the beach as
the water level was at low tide. We had to get the bow above
the high tide line. Afterwards we had a great time at the
Bowens' house. They said that the clam boats were a regular
feature on the river and they had very little boat traffic on
this section of the river, a fact confirmed by the sparsity of
boats we observed today. Only about ten boats total on the
river of which half were around the boat launching ramp at
Benedict. Mrs. Bowen provided us with steamed oysters and
sandwiches until we were picked up by Mrs. Vogt around
7:00 pm. J to K distance 0.95 miles.

DAY 2 OCTOBER 17, 1987

Point K: We agreed to meet at the Bowens' house at 10:30
am to resume the trip just before high tide. Mrs. Bowen came
down to the beach with Randi and Peggy to see us off. Our

launching was again a bit wobbly as we got adjusted to the canoe. Our aches from the last day of paddling probably resulted in slightly reduced speed. The morning had high fog which burned off around noon. Temperatures were in the low 60s when we started with much higher humidity, resulting in more perspiration than yesterday. Winds were from the north and southeast at about 5 knots maximum. We left the Bowen beach at 10:55 am and headed for Prison Point after rounding Kitt Point. K to L distance 1.7 miles.

Point L: Prison Point. We arrived at 11:35 am. This was a low marsh and rather attractive point. The air was calm and the tide was near high tide. We saw one small boat go up Battle Creek. We debated stopping at the Calverton site at Prison Point but decided to go on to Jack Bay due to anticipation of loss of calm seas (the weatherman predicted 15 knot winds). L to M distance 1.0 miles.

Point M: Marsh Adjacent to Jack Bay. We arrived at this marsh at 12:00 noon. This is a very attractive marsh. It has one duck blind at the tip of the point. Calm seas with no wind during this leg and maximum high tide. We explored some inlets in the marsh and departed at 12:10 pm after a brief rest. M to N distance 1.6 miles.

Point N: Point Near Parkers Wharf. Arrived at 12:50 pm. The eroding cliffs here exhibited interesting cross-bedding. Some shell was noted eroding out of portions of the cliffs. We left the cliffs at 1:00 after a short rest. Calm seas with north wind, two boats observed coming from Broomes Island area. N to O distance 2.0 miles.

Point O: Tip of Broomes Island. Arrived at 1:45 pm. This stretch was a lot of open water with fatigue setting in, pauses between pushes to reach the point of Broomes Island. We were aided by calm seas and no wind. The tide was going out at this point. We noted four speed boats during this crossing. The wake of one of the speed boats spilled a little water into the bow because I did not stop paddling as we headed into the wake. This was the only instance of swamping during the whole trip and the incident resulted in about a quart of water in the canoe. At Broomes Island we took a lunch break. We found that the tip of the island is an eroding headland or fastland which is composed of eroding marshland which is built up over former dry land surface. The area was also a site as evidenced by at least 40 fire-cracked rocks eroding from an area 50 feet to 150 feet east of the point. Peter also found two quartzite flakes along the south shore although the amount of larger cobbles was much less on the south shore. Along the north shore I found a quartzite stemmed point 70 feet from the tip of the island in an area with much fire-cracked rock. I also found a quartzite flake knife in the same area. The shoreline in this stretch was littered with a high density of cobbles extending in size up to 6 inches. This is a Late Archaic site eroding out of the marsh peat. The artifacts are retained for Jefferson Patterson Park and Museum. Peter stated he has also found points here during past walks along the shoreline...Point Farm looks very far away from this point. A portage of passage at Broomes Island would have been welcomed. A site form should be filled out for the Broomes Island site. Departed at 2:40 pm. O to P distance 2.8 miles.

Point P: Peterson Point—end of trip. Arrived at Peterson Point at 4:00 as predicted. The last leg was slowed by a slight wind in the second half of the leg. I felt sick from the orange [an over-ripe one eaten earlier] and was not nearly as effective as a bowman. We angled first toward the King's Reach house but then changed course for Peterson Point after I began to feel bad. Upon arriving at the beach, Randi greeted us. I used my truck to pull the canoe ashore this time. It was a very successful trip. The overall average speed—all in fair weather—was about 2.1 mph, of which a part was ebb tidal current assist. There is no doubt that experienced/fit paddlers of yore (Indian and Colonial) could have achieved higher speeds.

[Wayne Clark's trip log includes a graph showing times, distances, and average speeds during each leg.]

CHAPTER 2

Jug Bay to Hurricane Isabel

The last Patuxent River dugout trip—from Jefferson Patterson Park and Museum up St. Leonard's Creek and back to Jefferson Patterson—was advertised on a flier for the Park's Patuxent River Discovery Day on, May 7, 1988. *Rakiock* was to be paddled among a flotilla of regular canoes. As I wrote in 1990:

struggling to keep up with her sleek aluminum companions, *Rakiock* wallowed along behind, an ugly duckling but the star of the show. Our flotilla gathered at the site of the June 1814 naval engagement between US and British forces at the mouth of St. Leonard's Creek. Wayne [Clark] gave a short lecture.

A tentative plan to leave *Rakiock* at Jefferson Patterson on a semi-permanent loan was not pursued. Stored there for display until 1990 in a barn with a leaky roof, the canoe intermittently accumulated rainwater. No big problem, given all the preservatives I had sloshed on it back in early 1983. *Rakiock* was one of the standard attractions on the wagon tours offered to visitors.

Rakiock's only "commercial" deployment happened in the warm season of 1990, when the dugout was trucked from Jefferson Patterson to the marina of the Solomons Landing Condominiums to help celebrate (and promote) this development and its marina.

Rakiock's next deployment was at Jug Bay Wetland Sanctuary in Anne Arundel County. It was trucked there on a hot summer day in 1990 and remained there until about 1997. This park was practically on my way home from the National Research Lab in Washington so I stopped there often.

No long trips were taken from Jug Bay, but several deployments were recorded for special events. One undated flier, "Discovering Nature's Secrets," for a Patuxent River Discovery Day, directed visitors to the end of Otter Trail for dugout rides. This trail ends next to the waterfront, with land just a foot or so above sea level. The venue was freshwater tidal shallows. I could pull the canoe into the water on log rollers. My best launch site ever.

The Marsh Notes section in the Fall 1991 issue of the Jug Bay newsletter advertised me in Lectures-in-the-Field, on October 20. I teamed up with park director Chris Swarth to offer "From Buttons to Boats," from 2 to 5 p.m. The theme was making useful household items by hand from wood in your own yard.

October 26 was Childrens' Day at Jug Bay. I generally did not let others paddle *Rakiock* without me on board, usually in my stern seat. However on that day, as I wrote later, "the Jug Bay staff took turns ferrying young pretend Indians, freshly adorned with feather headdresses, and face paint."

I photographed a man named Andy Manele with three small "Indians" on board. Intentionally or not, he was paddling the dugout with the bow and stern reversed. Given the hull shape, it's not surprising the canoe can readily be paddled in either direction. My photo shows the leaves of marsh plants rising above the water. Could be spatterdock (cow or pond lily) or also tuckahoe (arrow arum) or pickerel weed, all common in Jug Bay and all once used for food or medicine by natives.

One of the trips turned into a bit of an adventure, as Shannon Loux and her charges got stuck in a strong tidal current and had to be towed back. Jug Bay is wonderful canoeing country, so long as one keeps track of the tides. One can get not only stuck as Shannon did trying to canoe back against the tide, but also literally get stuck, as I did, on a shallow mud bank or in a dense jungle of spatterdock—which may have been totally submerged and canoeable just an hour or two earlier. When that happens there is only one solution: One of the canoeists has to jump overboard and slog through the muddy bottom, pulling the dugout and hoping not to step on a snapping turtle. That works in a few inches of water. Even less than that, and you may have to wait for the tide to rise.

Regretfully, in 1991 I stopped keeping detailed logs of dugout deployments. I now had plenty of performance data and maybe just got bored keeping notes. Of course I did not foresee writing a book three decades later.

In 1997, *Rakiock* was trucked from Jug Bay to the beach just south of the mouth of Parkers Creek. This area on the western shore of the Chesapeake, known as Warrior's Rest, was preserved in 1996 by the Maryland Natural Heritage Program of the Department of Natural Resources via The Nature

Conservancy. It was purchased from the estate of the late Dr. Page Jett, a prominent Calvert physician. Since preservation, the 270-acre tract, which includes frontage along the Calvert Cliffs, has been managed by the American Chestnut Land Trust for education and research.

While never launched into Parkers Creek or on short Chesapeake Bay trips, the dugout starred in a group paddle along the Calvert Cliffs south to Flag Ponds Nature Park. *Rakiock* was to be accompanied by a flotilla of regular canoes and kayaks in a 6.5-mile trip sponsored by ACLT, Battle Creek Cypress Nature Center, Flag Ponds Nature Park and the Calvert Marine Museum Canoe Club. The flotilla set out at 9 a.m. on May 9, 1998, and the boaters paddled at their own rates. Dwight Williams, a Calvert County naturalist and park manager, made the trip by taking turns with me on a regular aluminum canoe and the dugout. Data on canoeing speeds and arrival time were not recorded. The outing was a success, and all boaters arrived at Flag Ponds safely.

The distance is comparable to the width of the Bay at Flag Ponds—so this was to be a test paddle for a Bay-crossing paddle, scheduled for June 13. I checked out a landing spot on the Eastern Shore—Taylor's Island Family Campground and Marina—a long drive, but the owners were okay with the concept. Several friends with motor boats volunteered to accompany me on the crossing, and I researched the issue of ship traffic and the inability of large vessels to stop on the proverbial dime. Crossing the navigation channel could be risky. The canoe was slow and looked like just a piece of driftwood. On June 13, the sea state was not ideal and the dugout crossing was scrapped. However a number of kayaks did cross on that day.

Due to logistical challenges and risk factors, no further dugout Bay crossings—although common in Indian times— were attempted. The dugout remained up on the wide Flag Ponds beach for the next fourteen years. When not actually deployed on the Bay, it was covered by a brown tarp. *Rakiock* was always parked above high tide next to the shallow embayment formed inside the southwards prograding spit. As these embayments became too shallow or turned into beach, the dugout was progressively over the years moved farther south. Numerous park visitors—adults and kids alike—got canoe rides, with many helping paddle.

My persona evolved from a late Colonial-era farmhand to an early Colonial fur trader. I acquired and always had along samples of black bear, wolf, mountain lion, woodland bison and, of course, a deerskin. The latter was big enough to cover the orange life preservers. When weather kept me on land, I would tell visitors about the wild animals that once lived here or within trading distance, species locally or regionally extirpated after European settlement.

Every now and then I would check up on *Rakiock*, maybe replacing a torn tarp, making sure the dugout was resting on logs, filling cracks with roofing tar, or brushing on more wood preservatives. Sometimes I would find that mice had built nests under the overturned canoe. On one occasion I was swarmed by angry ground bees whose nest I had disturbed. I ran between several bathing beauties to distract the bees, but the bees were not distracted.

In 2003, two decades after *Rakiock* was joy-ridden by young villains along the Alexandria waterfront, I once more feared my canoe was lost. But this time by Mother Nature herself, who had presented me with the tree in the first place and the same way—wind. It was Hurricane Isabel and its fierce winds and record storm surge. After the storm had passed, park staff saw no dugout. Days later, park manager Dwight Williams reported that Cub Scouts had found it lodged in the pioneer forest inland from the beach. The canoe was nearly filled with sand. Aside from a big but harmless scuff mark along one side, it had—unlike countless modern boats—escaped unharmed.

On its short wild ride inland, *Rakiock* had even dispatched a sapling or two.

CHAPTER 3

Seaworthy
at Calvert Marine Museum

Rakiock's last active deployments on tidewater began in 2012, when Calvert Marine Museum director Sherrod Sturrock invited me—and *Rakiock*—to demonstrate dugout use and give rides in their inner harbor. Paddling the dugout directly from Flag Ponds to the museum was possible, but I decided on a simpler, 10-mile journey by pickup. John Little, a neighbor and friend who managed the Flag Harbor Yacht Haven, offered to drive the canoe to the museum. Another neighbor helped me paddle the dugout the short distance north into the marina harbor and underneath their boat lift.

Operating the lift, John hoisted up the boat, cradled in a sling, and set it down into the back of his pickup. Being 15.25 feet long, the dugout protrudes aft beyond the tailgate on standard pickups. It was thus important to weigh down and strap the stern end of the canoe—now up near the pickup cab's rear window, and attach red flagging to the canoe end protruding from the back of the truck.

I sat in the passenger seat while John drove. We were out in the fast lane of MD 2–4, speeding south. Then the engine began to sputter.

We were running out of gas.

A vision of stalling and being rear-ended at high speed intruded my mind. The bow end of my canoe could smash through the window and bludgeon to death the driver of the vehicle behind us, or shove the other end of the dugout into the cab of our truck. A viral man-bites-dog type of story like this with my wooden relic would make global news. What would Indians or colonists have thought?

Lady Luck and John's nonchalant skill saved us from the jaws of crash. The turnoff to a gas station happened to be just ahead, and John sputtered and hiccupped the truck across the northbound lanes and alongside the pump.

Members of the Patuxent Small Craft Guild later helped park *Rakiock* along the outside wall of the Calvert Marine Museum's old-boat display building. We put the boat on wood blocks and covered it with a tarp.

For special events like Patuxent River Appreciation Day, *Rakiock* was launched repeatedly down into the museum's small lagoon, an effort that took several people. I paddled around, often with kids as passengers, but also alone, sometimes past the historical Drum Point lighthouse, moved there in the early 1970s, and a few hundred yards out into open water. Volunteers or staff helped passengers in and out of the dugout.

The behavior of children passengers once we got underway varied. Some were clearly nervous or even panicked, looking stiffly ahead, clutching the gunwales and wondering why their parents had made them do this. Others grabbed the gunwales and soon experimented with how much they could roll this thing back and forth. Parents back up on the dock were variously amused or alarmed.

On one launching, I took the tarp off and found the starboard gunwales punctuated with multiple carpenter bee holes. Gnawing through creosote and other preservatives apparently didn't bother these insects. I had never previously used wood filler in three decades of paddling and storage.

As 2014 moved along, I finally decided to hang it up and retire my dugout. I offered to donate it to the Calvert Marine Museum, a déjà vu from my 1970s plans. However, there were already two dugouts in the boat shed, with little remaining room. My offer was graciously rejected.

So I offered it to the Bayside History Museum on Fourth Avenue in North Beach, Calvert County. Grace Mary Brady, the founding force of the museum, lost no time accepting. She arranged for the truck, which arrived at 1:30 p.m. on August 4, 2015. *Rakiock* was hauled to the North Beach Department of Public Works for a one-off task that read: "make a cradle to secure canoe in place."

In its cradle, my hollowed-out, 600-pound *Rakiock* was hoisted up into the main floor of the museum, where it has been on display ever since, outfitted with the various furs and pelts, pole, paddles and bailing gourds.

Rakiock rests in context, against a backdrop panorama of local Indians in their small village at Colony Cove, painted by Deborah Watson.

If *Rakiock* could talk, he or she would be thankful that those orange, non-authentic life preservers are finally gone.

CHAPTER 4

As Time Goes By

Most events in this book took place thirty to forty years ago. Many monument trees that were then alive in the region are no more. The repurposed 6-feet in diameter rakìock, cut down in our community in 1986, has been previously discussed, as has the hollow reading nook rakìock which fell around 1990. The famous Annapolis Liberty Tree rakìock was downed by Hurricane Floyd in 1999. A severe thunderstorm in 2002 took out the famous Wye Oak on the Eastern Shore, the largest white oak in the United States. However, acorns previously collected from this monarch sprouted into seedlings that were distributed far and wide. I planted one in my yard in 1983. Thirty-three clones were also produced by Francis Gouin, professor emeritus of the University of Maryland Department of Horticulture and Landscape Architecture

The largest surviving American Chestnut in Maryland— namesake for the American Chestnut Land Trust—suffered from the blight, but actually broke off in a 2006 storm. The tree survives as a fair-sized stump sucker, its pollen spreading and DNA collected.

It's highly likely that none of the above trees had been planted by the hand of man. By contrast, in 1785, George Washington planted two rakìocks in the garden around Mount Vernon. He admired the species' majesty as well as its medicinal and other uses. While my 1983 field notes did not mention it, National Colonial Farm interpreter Wes Muller and I probably saw these two old trees on our Mount Vernon visit that year. Even today, one of them survives—238 years old in 2023. Another even older surviving raklock is Calvert County's champion tree in Flag Ponds Nature Park. This hollow giant has survived many storms, particularly tropical cyclones such as Hurricane Irene in 2011.

What if my 1980s dugout trips were repeated today? The Potomac, Mount Vernon, National Colonial Farm, Fort Washington, Gunston Hall, and historic Old Town Alexandria are much the same, but even more authentic with more visitors. At Mount Vernon, we've seen new archaeological discoveries. Then as now, the past experienced along the river is the late Colonial, early Federal past. Clad as farm hands of those times, we fit into this past as re-enactors.

While Natives had lived along these same banks for millennia, and our raklock dugout descended from their heritage, it was just as hard forty years ago as it would be today to reimagine the prehistoric Potomac, likely alive with dugouts. We found no native stone artifacts or lithic debris along the "natural" shores where we stopped. We only know from history and archaeology that the chief's Nacotchtank settlement was located along the east bank of the Anacostia, a bit north of its mouth—the site likely visible while we paddled just south of Hains Point. John Smith located but did not name Piscataway settlements near present National Harbor, and another near Mockley Point. Assoameck was about where Belle Haven is today. Another chief village of the Moyaons—the name recorded by Smith for the Piscataways—once lay a bit south of present National Colonial Farm. The name survives as Moyaone Reserve, a large-lot development near the Farm, designed to preserve the historical viewshed from Mount Vernon.

Paddling upriver today, we would see a Woodrow Wilson Bridge much changed from the one we canoed under in 1984, which was built in 1961. From 2001 to 2008 the bridge was rebuilt to accommodate the ever-increasing vehicle traffic. While the old bridge carried six lanes, the new one has twelve, and is 20 feet higher. Across the Potomac from Cameron Bay, National Harbor dominates the view today. Paddling toward the bridge from the south in 1984, we could not have known that the 300-acre Salubria Plantation was being sold that same year. The proposed mega development PortAmerica was approved in 1988. The current population of National Harbor is 5,500, mostly living in riverside condos. Today we would encounter far more boats, mostly from the new marina.

Approaching Old Town Alexandria, today we no longer would see Robinson Terminal North and the warehouse, both demolished in 2020 as part of RiverRenew. This project includes a new tunnel system to prevent Alexandria sewage and stormwater overflow from reaching the Potomac.

Overall water quality along our 1983–1984 paddles was not as good as it is today. The river we crossed forty years ago to Mount Vernon was green with algae. But even then the river was already light years cleaner than fifty years earlier, when raw sewage flowed into a stinking waterway. Founded in 1940, the Interstate Commission on the Potomac River Basin master-minded progressive river restoration. Improvements continued intermittently, especially in the 1970s and 1980s. The first Potomac Riverfest in 1980 and the one we attended in 1984 celebrated returning the tidal Potomac near DC to a cleaner, fishable river. While we took scant notice while paddling past it on 10 June, 1984, the Blue Plains Advanced Wastewater

Treatment Plant now is the largest plant of its kind in the world. Three hundred million gallons of water are cleaned on an average day, with a billion on peak days.

The unseen ecosystem below our dugout also has changed since, with invasive snakeheads, introduced around 2002, and two large catfish—flathead and blue—impacting native species. We saw no hydrilla in 1984, although it was first noted in 1982.

In 1984 we tied up *Rakiock* to a floating dock under a ramp owned by the Department of Education. This dock was removed many years ago. On that June 1 voyage, round noon, we paddled past two large vessels. One was the salvage and rescue tender, USS *Hoist*, which would be scrapped in 2007, and the other, the schooner *Alexandria*—whose rechristening we had observed—which would sink in a storm off Cape Hatteras in 1996. The Red Cross Alexandria Waterfront Festival, first held in 1980, lasted for twenty-eight years but is no more. The festival required too much organizational work for the revenue generated.

Along the tidal Patuxent, Jug Bay, Patuxent River Park, King's Landing, and Jefferson Patterson Park and Museum are better than ever, with more visitors, more events and at Jefferson Patterson more and new archeological discoveries and exhibits. Judge Perry Bowen and his wife, Virginia, moved to Virginia from their historic house, where Wayne Clark and I had interrupted our voyage for the night. The judge and Virginia have passed on.

Taking the same trip from King's Landing to Jefferson Patterson today, we would see much the same shoreline sights, such as the historic Sheridan Point plantation house. Passing

Broomes Island, we would remember that Bernie Fowler, born and raised there, had passed in 2021. The Broomes Island oyster packing plant also is no more. Stoney's Seafood House, which had opened in 1989, two years after we paddled past it, has been gone since 2020.

We would still complain today about the unsightly, coal-fired Chalk Point Generating Station—and now also about its contribution to climate change. And about the overhead transmission lines and the Maryland Route 231 Benedict Bridge, built in 1952. We would see more boats, but would only complain when passed by a roaring jet boat. The dugout would ride its wake with little or no water splashed in, so our complaints would focus more on noise and those carbon footprints scarcely thought of in 1987.

Although there's still more suburban development along the Patuxent today, there is also more riverside farm and forestland preserved, especially on the Calvert County side. We would see no tobacco fields: Maryland's tobacco buyout program began in 1999, but when we paddled in October 1987, the tobacco had long been harvested and was curing inside barns.

Evidence of ancient Native Americans can still be found on some beaches. Despite population growth, much of the shoreline and its marshes preserve a sense of this segment of the tidal river's natural history as experienced by the Patuxent Indians four centuries ago. Captain John Smith had shown some seventeen native hamlets along the tidal Patuxent in 1608, of which he marked ten in the 21-mile stretch we paddled. Spelled phonetically by Smith and modern linguists, the names are all tongue twisters to us today. There were, west of the river: Pocotamough, roughly across from King's

Landing; then southwards after a gap of six miles, beyond God's Grace Point, Macocanaco, Wasupokent, Acquaskack, Wasinacus, and Acquintanacacsuck. The latter was a chief's village and lay opposite Broomes Island. Wascocup, just south of Hallowing Point, was followed by Onuatuck (between Prison Point and Kitts Point) and then, on the south bank of outer Battle Creek, the chief's village, Pawtuxunt, which gave the river its current name. Quomocac lay several miles south of Pawtuxunt and within the present area of the Jefferson Patterson Park and Museum.

The seventeen-plus native villages along the tidal Patuxent had no outhouses, septic tanks or wastewater treatment plants. No evidence for latrines has been found. Calls of nature were evidently answered in selected woods, not too close to the hamlets. It's very likely that many of the villages were even within 1,000 feet of tidewater, the so-called Maryland Critical Area. Few tidal rivers were as densely populated as the Patuxent, yet John Smith commented on its "infinite schools of fish, more than elsewhere."

As late as 1859, author James Hungerford used the name Clearwater River for the Patuxent. At a time of maximum agriculture and eroding fields, he nevertheless gushed that "of all the rivers flowing into the Bay, there is not one which excels the Clearwater in the purity of its waters and the variety and loveliness of its picturesque scenery...So transparent are the waters far from its shore you may see, in the openings of the seaweed forest on its bottom the flashing sides of the finny tribes as they glide over the pearly sands."

By 1987, the comparative condition of the Patuxent and Potomac had reversed. On our canoe trip we were not focused on looking into the water, but would have noticed, and commented on, seaweed forests and finny tribes with flashing sides. The absence of any mention in archaeologist Wayne Clark's comprehensive log speaks volumes about lack of water clarity on our voyage.

THE LEGACY OF BERNIE FOWLER

In 1987, former Broomes Island waterman C. Bernard Fowler, by then a Calvert County political leader, had observed to his sorrow that the Patuxent River had become so murky he could no longer see his feet while wading waist deep. Fowler went on to lead the campaign to clean up the river, which had become loaded with phosphorus and nitrogen compounds from new upriver sewage plants, private septic systems and abundant commercial fertilizer from farms and gardens. Inspired by folk singer and environmentalist Tom Wisner, Bernie Fowler began an annual Patuxent River group wade-in, held on the second Sunday of each June. Clarity would now be measured by the Sneaker Index, measuring the depth at which Bernie no longer could see his sneakers while looking down into the water. These annual wade-ins—first from Broomes Island and starting in 2010 from Jefferson Patterson Park—have become annual media and social events.

The first wade-in was held in 1988, the year after our dugout voyage. In that year the index was abysmally low: only 10 inches. Next year it was even worse: 8 inches. But then, thanks to better waste-water treatment and other factors, the index improved, reaching 44.5 inches by 1997. Then, after several years with no major changes, a general decline followed. The index fell to as low as 21 inches. The index measured 47 inches in 2019 but would plunge again, to 39 inches in 2022 and then 30 inches in 2023.

Paddling the dugout today, we would again see some seaweed forests and finny tribes. While the Sneaker Index for 1608 and 1850 can only be guessed, it was probably even higher than the five feet Bernie recalled from the 1950s and 1960s, an achievable goal.

The Chesapeake Bay Critical Area Protection Act passed in 1984, the year we paddled the dugout on the Potomac. To forestall and maybe reverse declining water quality, further shoreline development throughout Maryland tidewaters became very restricted. (By then, cleaning the tidal Potomac near Washington DC was well ahead of other areas, including the Patuxent.) Now, septic systems must be replaced by costly nitrogen-removal technology. Shoreline forest buffers are to be preserved and expanded. The act was implemented by

Calvert County after a contentious public hearing, vehemently opposed by owners of waterfront land. By coincidence, that hearing closely coincided with our fall 1987 dugout trip down the Patuxent.

If the 1998 dugout trip from Parkers Creek to Flag Ponds was repeated today, paddlers would scarcely note changes. The six developments along the cliffs—Scientists Cliffs, Governor Run, Kenwood Beach, Western Shores, Long Beach and Flag Ponds—have not expanded, with only a few new or enlarged cliff-top houses. Except for the 800-foot section of cliff at Governor Run, no additional land has been preserved along the Calvert Cliffs. At Flag Ponds the spit has migrated further south, creating new land, while the now forested northern part of the spit is being eroded. Restoration of oyster reefs and sturgeon remain challenging dreams. However, overall Chesapeake Bay health is better today.

Conclusion

Unlike those ancient Saxon pagans, I don't worship trees. But I venerate them, particularly large and old ones. Unlike Saint Boniface I don't think my wonderful tulip poplar, aka rakìock, was toppled by a divinity. But like that ancient British monk, I did make things from the wood—albeit not to score religious points. To be sure, I wanted to make mementos—along the lines of things made from the Wye Oak and the Liberty Tree, only much bigger one-offs.

The 1980 gust of wind led me to begin liberating large rakìock creations—a kubbestol, a dugout canoe and later, from other trees, a giant round table and a reading nook. And that led to this book. It was all educational and healthy good fun, connecting me to trees, wood, ancient traditions and historical reenactment. Maybe the many who over the years rode in or viewed *Rakìock*, or read this book, come away with a little more sense of human and natural history, primitive technology and travel, prehistory, wonderful waterfront scenery and, not least, the challenge of environmental restoration.

Sitting and paddling many hours in my primitive and humble craft, sometimes rolling gently like a miniature icebreaker, I had plenty of time to ponder those matters. After floating in a tree which once graced our yard, I came home to relax inside that same tree, albeit upside down, with my head pointed into the ground.

PHOTO CREDITS

Foreword

"Poplar at Flag Ponds -2022."
Photo by Suzanne Shelden. - page vi

Table of Contents

"Trick of the eye—Flag Ponds poplar. - 2022"
Photo by Suzanne Shelden. - page viii

Introduction

"Fallen tree at Vogt house."
Photo courtesy of the Vogt family. - page xii

PART I

"Troll Pulpit at Battle Creek Nature Center."
Photo by Suzanne Shelden. - page xvi

Chapter 1

"Peter and standing rakìock."
Photo courtesy of the Vogt family. - page xviii

"Tulip tree flowers."
Image from Donald Peattie's *Natural History*. - page 3

Chapter 2

"1907 photo of the Liberty Tree on the campus of St. John's
College. Inside are the family of Mayor Gordon H. Claude
plus five other Annapolis girls."
Collection of the Maryland State Archives. - page 4

"Flag Ponds tulip poplar in March, 2015,
with a quartet of visitors inside."
Photo by Karyn Molines, Calvert Country Natural Resources
Division; https://www.flickr.com/x/t/0000009/photos/
kmolines/22102111061/
- page 8

"Daniel Boone's Beech, 1890."
Postcard image courtesy of the Washington County,
Tennessee, Archives and Betty Jane Hylton.
- page 13

Chapter 4

"The enormous root ball became a jungle gym for our
young son Jason (top) and his friends."
Photo courtesy of the Vogt family. - page 23

Chapter 5

"Kubbestol."
Photo courtesy of the Vogt family.- page 26

"Kubbestol with lip."
Photo courtesy of the Vogt family. - page 27

"Kubbestol with child."
Photo courtesy of the Vogt family. - page 33

Chapter 6

"Many rakiock creations."
Photo courtesy of the Vogt family. - page 34

"Scientist Cliffs neighbors with giant from which 'cookie' was
cut. Note how old open-field trees lose lower limbs and the
scars are healed by new bark to produce those knobs."
Photo by Peter Vogt. - page 39

"Peter with 'cookie'. "
Photo courtesy of the Vogt family. - page 41

" 'Cookie' becomes a table."
Photo courtesy of the Vogt family. - page 42

Chapter 7

"Two grown men—the author, right, and school friend
Wilbur Kendig—stand comfortably inside the emerging Troll
Pulpit, created between 1997 and 2012."
Photo courtesy of Andy Brown. - page 44

"American Chestnut Land Trust's fallen giant."
Photo by Peter Vogt. - page 46/47

"Duwane 'Duey' Rager about to enter the Reading Nook in
the Battle Creek Cypress Swamp Nature Center."
2022 photo by Duwane Rager. - page 54

Chapter 8

"Bry. Theodor De. Engraver, and John White. *How They
Catch Fish.* Native men and women fishing from a dugout
canoe. Virginia, 1590." Photograph retrieved from the Library
of Congress. https://www.loc.gov/item/2001696969/.
- page 56/57

"Bry. Theodor De. Engraver, and John White. *How they Build
Boats.*" Native men making a dugout canoe. Virginia, 1590."
Photograph retrieved from the Library of Congress.
https://www.loc.gov/item/2001696968/. - page 60

"Dugout canoes among other boats on display at Calvert
Marine Museum. The old wreck dominating the scene is a
multilog hull. To its left are a Calvert Marine Museum replica
of a 19th century one-log rakìock 'punt' dugout that I helped
make there. To its left is a replica like *Rakiock* but built with
power tools and charred to resemble native method."
Photo of display at Calvert Marine Museum. - page 63

"*Rakiock* Canoe with mural by Deborah Watson."
Courtesy of Grace Mary Brady, Bayside History Museum.
- page 64/65

Chapter 9

"A storm deposited a proper log in my backyard."
Photo courtesy of the Vogt family. - page 66

"Roughing out the dugout interior with an axe."
Photo courtesy of the Vogt family. - page 69

"Carved seat shields eye bolt nut from view."
Photo by Peter Vogt. - page 70

"Racheting the dugout out of the woods ."
Photo courtesy of the Vogt family. - page 74

"Jason Vogt paddles out of the woods."
Photo by Peter Vogt. - page 75

"Painting the inside with a turpentine-linseed mixture."
Photo courtesy of the Vogt family. - page 76

Part II

"*Rakiock's* journey up and down the Potomac River."
Drawing by Peter Vogt. - page 78/79

Chapter 1

"The view of Mount Vernon from the National Colonial
Farm. It was established in 1958 by the Accokeek Foundation
and demonstrates 18th century farming techniques to
visitors." 2012 Photo courtesy of Preservation Maryland,
Wikimedia Commons. - page 80

"By wagon to the river."
Photo courtesy of Peter Vogt. - page 83

"Into the river." Photo courtesy of Peter Vogt and National
Colonial Farm staff. - page 85

Chapter 1

"Two passenger Scouts could not capsize *Rakiock*."
Photo courtesy of Peter Vogt." - page 119

"Archaeologist Wayne Clark, bow canoeist, kept a careful log
of *Rakiock*'s performance and voyage."
Photo courtesy of Peter Vogt." - page 121

"Peter approaching Sheridan Point Farm."
Photo by Wayne Clark." - page 126

Chapter 2

"Peter and Randi Vogt at Solomons Landing."
Photo courtesy of the Vogt family. - page 130/131

"Jug Bay's Andy Manele ferries young pretend Indians
on Children's Day at Jug Bay."
Photo by Peter Vogt. - page 132

Chapter 3

"Peter paddling a wee blonde girl around the Calvert Marine
Museum inner harbor. She tried hard to rock the boat as soon
as we got away from the dock and her parents."
Photo by Calvert Marine Museum volunteer Bob Hall.
- page 139

"*Rakiock* in retirement at Bayside History Museum."
Courtesy Grace Mary Brady, Bayside History Museum.
- page 141

Chapter. 4

"Tom Wisner and Bernie Fowler
with Gov. Martin O'Malley and Betty Fowler."
Photo by William Lambrecht.
- page 148

SOURCES

Brewington, M.V. *Chesapeake Bay Log Canoes and Bugeyes.* Cambridge, Maryland: Cornell Maritime Press, 1963.

Brown, Russel C., and Brown, Melvin L. *Woody Plants of Maryland.* Baltimore, Maryland: Port City Press, 1972.

Brush, Grace S., Lenk, C., and Smith, J. "The Natural Forests of Maryland: An Explanation of the Vegetation Map of Maryland." *Ecological Monographs* 50 (1980): 77-92.

Clark, Wayne. "The Patuxent Indians at the Dawn of History," *Calvert Historian*, 22 (1996): 6-20.

Clark, Wayne. "Official Log: Patuxent River Log Canoe Experimental Trip Number One, October 16 and 17, 1987." Unpublished, 1988.

Clark, Wayne. "Algonquian Culture of the Delaware and Susquehanna River Drainages: A Migration Model." COMCAR Project No. 09-26 No. DEWA 2010 A, 2019.

Curran, H.M. "The Forests of Calvert County," in *Calvert County, Maryland Geological Survey*, pp. 214-222. Baltimore, Maryland: The Johns Hopkins Press, 1907.

Hamilton, Sherry. "Reed constructs authentic dugout canoe." *Gloucester Mathews Gazette-Journal* 86 (2023): 5.

Hariot, Thomas. *The First Plantation of Virginia.* 1588 and 1590. Reprint. London: Bernard Quaritch, 1893.

Hulton, Paul. America 1585: *The Complete Drawings of John White.* Chapel Hill, North Carolina: The University of North Carolina Press, 1984.

Hungerford, James. *The Old Plantation and What I Gathered There in an Autumn Month.* New York: Harper and Bros., 1859.

Lavish, Al, and Surgent, George. *Early Chesapeake Single-Log Canoes.* Solomons, Maryland: Calvert Marine Museum, 1984.

Lopes. Jane. "Transportation Used Many Years Ago To Travel The Nemasket Is Recreated." *Middleboro Gazette*, 25, 1995.

Madsen, Jan S. Stammebaade: *Dugouts, Einbaume, Pirogues, Vikingeskibshallen I Roskilde.* Denmark: 1985.

Mariner, Kirk, "The Methodist: Parson Thomas and His Canoe," *Chesapeake Bay Magazine*, August 1987, pp. 40-41.

Michener, James. *Chesapeake*. New York: Random House, 1978.

Peattie, Donald Culross. *A Natural History of Trees of Eastern and Central North America*. New York: Bonanza Books (Crown Publishers), 1966.

Royse, Lisa. "Preserving the Past for the Future: Conservation of the Dugout Canoe." *The Mariners' Museum Journal* 15 -16 (1989): 22-24.

Rountree, Helen. *Pocahontas's People*. Norman, Oklahoma: University of Oklahoma Press, 1990.

Rountree, Helen C.; Clark, Wayne E.; and Mountford, Kent. *John Smith's Chesapeake Voyages 1607-1609*. Charlottesville: University of Virginia Press, 2007.

Semmes, Raphael. *Captains and Mariners of Early Maryland*. Baltimore, Maryland: Johns Hopkins Press, 1937

Shomette, Donald. "A Sub-Surface Radar Exploration of Lake Phelps, North Carolina: September, 1992." Raleigh, North Carolina: North Carolina Division of Archives and History, 1993.

Steen, Albert. *Stoler i Norge [Norwegian Chairs]*. Oslo: C. Huitfeldt, 1975.

Sutton, Ann, and Sutton, Myron. *Eastern Forests*. The Audubon Society Nature Guides. New York, NY: Alfred A. Knopf, 1985.

Tilp, Frederick. *This Was Potomac River*. Self-published, 1978.

Treuer, Anton. *Atlas of Indian Nations*. Washington, DC: National Geographic Society, 2013.

Tyler, Lyon G., ed. *Narratives of Early Virginia*. New York: Charles Scribner and Sons, 1907.

Vogt, Peter. "Tulip Poplars: Friendly Green Giants?" In *The Peeper*, Newsletter of the Battle Creek Nature Education Society, 25(2): 1, 6-7.

Vogt, Peter. *Tourmaline's Quest*. Santa Barbara, CA: Santa Barbara Books, 2020.

About the Author

A marine geoscientist by career, Dr. Peter R. Vogt was educated at Caltech and the University of Wisconsin. Undergrad wanderlust led him to climb Mexico's high volcanoes, travel overland along the Andes, work for a mining company in the blistering Arizona desert and get his first taste of oceanography on a Scripps Pacific expedition. As a new Wisconsin grad student, he first helped a professor in Antarctica (1962–'63). Next year Vogt enlisted as research watch-stander on a seven-month Woods Hole expedition to the Indian Ocean, and collected thesis data on two icebreaker cruises (1965 and 1966) tailed or buzzed by the USSR to the European Arctic.

Vogt's career as Navy civilian scientist spanned almost four decades: 1967–'75 at the Naval Oceanographic Office and 1975–2004 at the Naval Research Laboratory. Besides applied research, he participated in basic science research on many US vessels and some foreign ones (Norwegian, French, Soviet and Russian). Most of the cruises investigated the bottom and sub-bottom of poorly known deep ocean venues-the North and South Atlantic, Arctic, Pacific, and Indian oceans. Some memorable deployments were on the US nuclear submarine NR-1 (1999), two 10-hour dives (1998) in the Russian MIR submersibles in which director James Cameron explored the *Titanic* wreck for his movie, *Titanic*; and a stint as co-chief scientist on the Glomar Challenger (1975).

Vogt has published more than 150 research papers, cited to date nearly 10,000 times. The topics included plate tectonics, rock magnetization, submarine volcanoes and massive landslides, and methane seeps. Best known outside the research world is the popular global chart This Dynamic Planet, a collaboration with the Smithsonian and US Geological Survey.

After moving to Calvert County, Maryland (1969) with his wife Randi and toddler son, Vogt focused his avocational interests on local natural history, especially the Calvert Cliffs, and preserving natural and farm land from suburban sprawl. In retirement he has continued his woodcraft hobby (locally found wood) and begun writing science-based science fiction (Amazon Kindle) for readers from school age to adult. Four stories are set in Southern Maryland, especially the Calvert Cliffs and Chesapeake Bay, and two books take readers into the prehistoric past.

Acknowledgments

I thank Jason Vogt, Wayne Clark and especially Fran Armstrong for reading, critiquing and correcting early drafts of this book. Thanks also to the many folks who have helped me paddle and/or launch or drag back out of the water or load on and off trucks, and all those I met on those dugout travels. And absolutely also my wife who tolerated those hobbies and put up with my hoards of wood under unsightly tarps or cluttering our basement.

APPENDIX ONE

Instructions for transporting Rakìock, page one.

(1)

IMPORTANT — PLEASE READ

TIPS ON HANDLING & MOVING DUGOUT

SO I CAN SLEEP SOUNDLY —
HAVE SEVERAL 100'S HRS INVESTED !!

TULIP POPLAR WOOD IS RELATIVELY
WEAK + SOFT + THE DUGOUT IS <u>HEAVY</u>.
if too MUCH OF ITS WEIGHT IS
SUPPORTED ON a SHARP EDGE OR
POINT IT WILL GOUGE HOLES IN
THE WOOD. THESE WILL "CATCH" ON
THINGS LATER. + WILL ROT FIRST.

DO NOT PULL BOAT ACROSS,
OR SUPPORT IT ON, OR DROP IT
ON, A SHARP WOOD OR METAL EDGE.

← NO

NO   OK

ALSO <u>PLEASE</u> DO NOT PUT A CHAIN, CABLE,
OR GRAPPLING (LOGGERS) HOOK AROUND
OR INTO IT AT ANY TIME.

I HAVE FOUND <u>ONE</u> PERSON
CAN MOVE IT <u>SLOWLY</u> WITH a FLAT
STRONG BOARD (6FT 2×6, SAY) +
A FULCRUM LOG, BY PUSHING DOWN
(LEVERING) + TURNING AT SAME TIME.

PUSH DOWN
+ TURN,
MOVING
$ END OF BOAT   short log.

OVER

THIS METHOD IS GOOD FOR TURNING

APPENDIX ONE

Instructions for transporting Rakìock, page two.

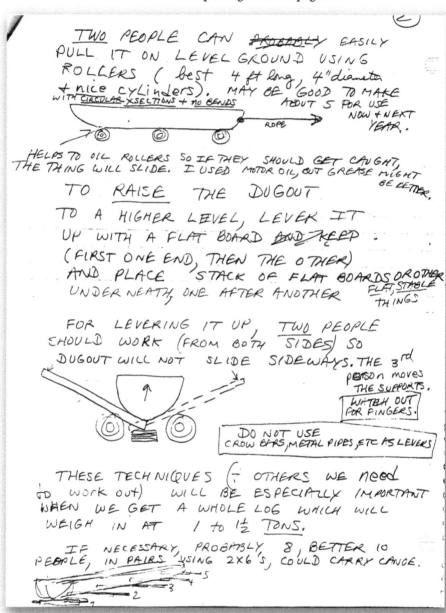

TWO PEOPLE CAN ~~PROBABLY~~ EASILY PULL IT ON LEVEL GROUND USING ROLLERS (best 4 ft long, 4" diameter + nice cyLinders). WITH CIRCULAR XSECTIONS + no BENDS MAY BE GOOD TO MAKE ABOUT 5 FOR USE NOW + NEXT YEAR.

ROPE

HELPS TO OIL ROLLERS SO IF THEY SHOULD GET CAUGHT, THE THING WILL SLIDE. I USED MOTOR OIL, BUT GREASE MIGHT BE BETTER.

TO RAISE THE DUGOUT TO A HIGHER LEVEL, LEVER IT UP WITH A FLAT BOARD ~~AND KEEP~~ (FIRST ONE END, THEN THE OTHER) AND PLACE STACK OF FLAT BOARDS OR OTHER FLAT, STABLE THINGS UNDERNEATH, ONE AFTER ANOTHER

FOR LEVERING IT UP, TWO PEOPLE SHOULD WORK (FROM BOTH SIDES) SO DUGOUT WILL NOT SLIDE SIDEWAYS. THE 3rd person moves THE SUPPORTS.

WATCH OUT FOR FINGERS.

DO NOT USE CROW BARS, METAL PIPES, ETC AS LEVERS

THESE TECHNIQUES (+ OTHERS WE need to work out) WILL BE ESPECIALLY IMPORTANT WHEN WE GET A WHOLE LOG WHICH WILL WEIGH IN AT 1 to 1½ TONS.

IF NECESSARY, PROBABLY 8, BETTER 10 PEOPLE, IN PAIRS USING 2x6's, COULD CARRY CANOE.

APPENDIX ONE

Instructions for transporting *Rakiock*, page three.

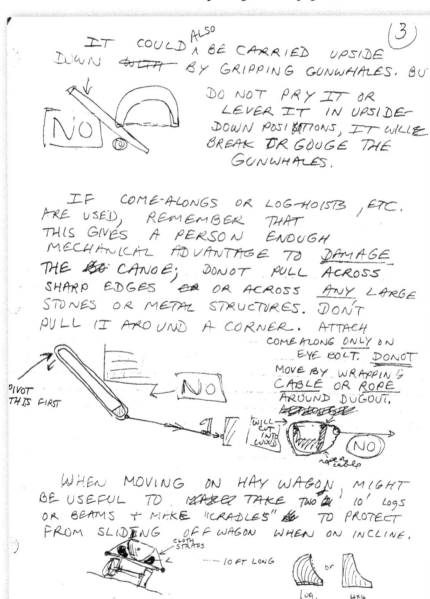

IT COULD ALSO BE CARRIED UPSIDE DOWN ~~WITH~~ BY GRIPPING GUNWHALES. BU'

NO

DO NOT PRY IT OR LEVER IT IN UPSIDE DOWN POSI~~T~~TIONS, IT WILLE BREAK ~~OR~~ GOUGE THE GUNWHALES.

IF COME-ALONGS OR LOG-HOISTS, ETC. ARE USED, REMEMBER THAT THIS GIVES A PERSON ENOUGH MECHANICAL ADVANTAGE TO DAMAGE THE ~~BE~~ CANOE; DONOT PULL ACROSS SHARP EDGES ~~OR~~ OR ACROSS ANY LARGE STONES OR METAL STRUCTURES. DON'T PULL IT AROUND A CORNER. ATTACH COME-ALONG ONLY ON EYE BOLT. DONOT MOVE BY WRAPPING CABLE OR ROPE AROUND DUGOUT.

PIVOT THIS FIRST

NO

WILL CUT INTO WOOD

NO

WHEN MOVING ON HAY WAGON, MIGHT BE USEFUL TO ~~MAKE~~ TAKE TWO ~~O~~ 10' LOGS OR BEAMS + MAKE "CRADLES" ~~#~~ TO PROTECT FROM SLIDING OFF WAGON WHEN ON INCLINE.

CLOTH STRAPS

---- 10 FT LONG

log.

or

4×4

APPENDIX TWO

Sketched plans for *Raklock,* page one

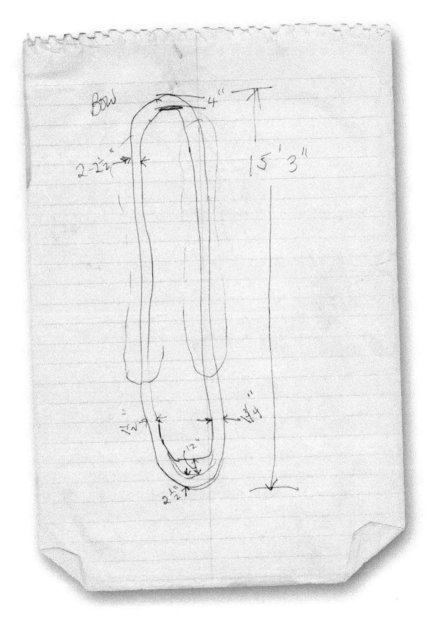

APPENDIX TWO

Sketched plans for *Raklock,* page two

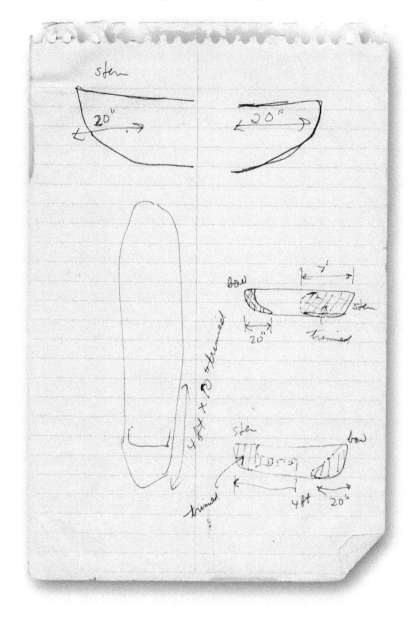

APPENDIX TWO

Sketched plans for *Rakíock,* page three

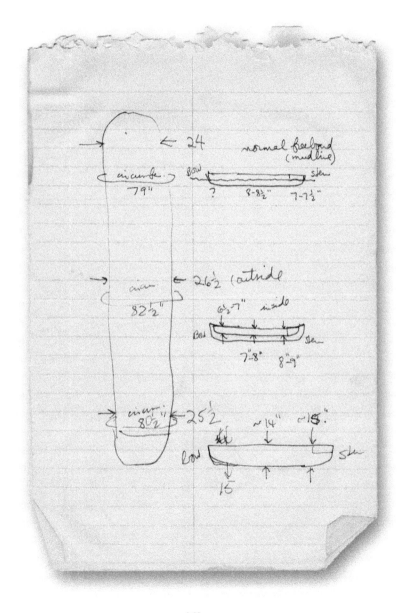

APPENDIX THREE

Samples from original *Rakiock* logs

used colonial bricks

8 bricks = 50 lbs
8 =
9 =
9 =
9 =
9 =

――――――――――
300 lbs.

+ 8 bricks = 40 lbs
――――――――――
total bricks = 340 lbs.

TRIAL load 1   4 small pumpkins
2 wood paddles   375
2 sheepskins   165
340 lbs. bricks   115
Brenda ?   (655)
P. 165 lbs
――――――――――

2nd Trial Percy = 200 lbs
total 860
3rd PV + everything + 315 lbs corn

APPENDIX TWO

Samples from original *Raktock* logs

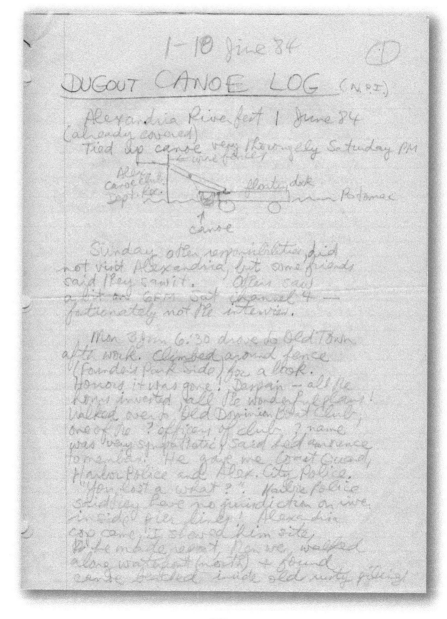

APPENDIX TWO

Samples from original *Raklock* logs

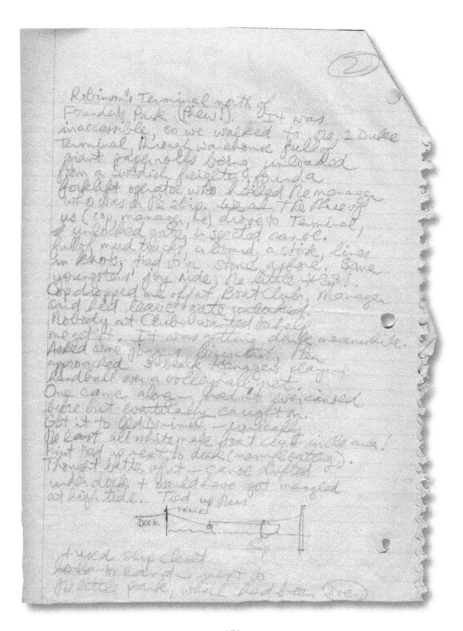

APPENDIX TWO

Samples from original *Rakiock* logs

Gangplank Marina → National Col F.

Wednesday June 20

Farm 9 AM. John Bros and brought Susan
Rhoads 9:30 drove us to Gangplank

11:00	lv. Gangplank Marina
11:40	Haines Pt.
13:10	NRL Pier
14:10	Woodrow Wilson Pts Bridge
16:30	Fort Washington
17:45	Colonial Farm
17:45–18:45	ratchet up + secure canoe.

weather warm, relatively low humidity, high 80's.
scattered variable altocumulus + cirrus.
winds 0 to 10 kts north backing to west.
variable, light sea. Relatively
few boats on river. DC Harbor patrol
cruised up to check us for life preservers.
Haines pt. — A. Wash Post or Times
native writer took flier, said he
would get in touch.

$$\begin{array}{r} 2.4 \\ 5\,\overline{)12} \\ 10 \\ \overline{20} \end{array}$$

$$\begin{array}{r} 6.75 \\ 1.7 \\ 7\,\overline{)12} \\ 7 \\ \overline{50} \end{array}$$

APPENDIX TWO

Samples from original *Rakiock* logs

18 Oct 84

1230–1325   50-55 minutes

National Colonial Farm — Mt. Vernon Pier.
canoed alone. Slight ~~edge~~ incoming
to high tide. 3"-6" waves from north,
light winds. partly cloudy to sunny, hazy,
warm (70's), water low 60's. First ~~of~~
trip alone.

[margin: 1 canoe]

1415 – 1620   Mt. Vernon to
Gunston Hall "Canal" mouth. ~~to~~
canoed with ③ canoeists (Greg Vink, John Trieul,
+ myself). Calm, glassy water to slightly
rippled ~~no~~ weak to moderate south-going
tidal current. Hydrilla patches, driftwood,
some with gulls on it. ~~Only~~ Only 3 boats
passed. Somewhat tippier with 3, but no mishaps. No water taken.

[margin: 3 canoeists]

1630–1730   explored "canal", beautiful
smell wooded inlet, autumn leaves.

Got canoe stuck on submerged log twice,
had to get into water to stand on log and pull it
off. Greg broke his paddle prying. Tied up canoe
just inside cove, next to bank. 1 mile walk
to Gunston Hall through fine old oak forest.
They were preparing for "Regents" dinner + were
not happy to see us.

Milton Keynes UK
Ingram Content Group UK Ltd.
UKHW020724011223
433542UK00011B/122